# Introduction to Quantum Information Processing

This book introduces quantum computing and quantum communications at the undergraduate level for students in the physical sciences, engineering, and mathematics, assuming no prior knowledge of quantum mechanics. It is a self-contained guide assuming only that the reader is familiar with physics at the college level, calculus (up to and including ordinary differential equations), and some matrix algebra and complex numbers. The book brings the reader up to speed with fundamental concepts in quantum information processing and builds a working knowledge of basic quantum algorithms, quantum error correction, and quantum communication protocols. It covers various models of quantum computation and some of the most common physical realizations of qubits. There is a good number of practice problems and exercises that rely on computer programming with the Qiskit language. The book can be used to prepare students of physics, mathematics, electrical and computer engineering, computer science, optics and photonics, and mathematics for internships and research experiences in quantum information processing, both during and after their undergraduate studies. It also helps those who plan to apply to graduate school and do research in this area.

**Key Features:**

- Self-contained coverage of quantum computing and quantum communications, from the motivations to the fundamentals and applications, including key concepts and contemporary topics.
- Contains practice problems and exercises, including some that require programming in Qiskit (the python-based, high-level language for programming quantum computers, adopted by nearly all quantum hardware providers and completely open source).
- Very light background knowledge assumed, making this book accessible to a variety of majors in the natural sciences, engineering, and mathematics.

**Eduardo R. Mucciolo** is a Professor of Physics at the University of Central Florida, USA, which he joined in 2004 after faculty positions at PUC in Rio de Janeiro in Brazil, and at Duke University in the USA. He has B.S. and M.S. degrees in physics from the University of Sao Paulo in Brazil and a Ph.D. in physics from the Massachusetts Institute of Technology in the USA. His areas of research are theoretical condensed matter physics and quantum information processing. He is a fellow of the American Physical Society.

# Introduction to Quantum Information Processing

Eduardo R. Mucciolo

CRC Press
Taylor & Francis Group
Boca Raton  London  New York

CRC Press is an imprint of the
Taylor & Francis Group, an **Informa** business

Designed cover image: Shutterstock

First edition published 2026
by CRC Press
2385 NW Executive Center Drive, Suite 320, Boca Raton FL 33431

and by CRC Press
4 Park Square, Milton Park, Abingdon, Oxon, OX14 4RN

CRC Press is an imprint of Taylor & Francis Group, LLC

ISBN: 978-1-032-77863-1 (hbk)
ISBN: 978-1-032-77589-0 (pbk)
ISBN: 978-1-003-48512-4 (ebk)

DOI: 10.1201/ 9781003485124

Typeset in Latin Modern Font
by KnowledgeWorks Global Ltd.

Publisher's note: This book has been prepared from camera-ready copy provided by the authors.

*For my girls*

# Contents

# Preface

This book grew from lecture notes developed for the junior-level course *Quantum Information Processing*, which has been offered regularly to computer science, mathematics, physical sciences, and engineering majors at the University of Central Florida since 2022. The course aims to introduce basis concepts, ideas, and methods in quantum information to undergraduates who had no previous exposition to quantum mechanics. Thus the prerequisites for the course are rather minimal: Physics I and II (calculus-based mechanics, electricity, magnetism, and optics), and Calculus with Analytic Geometry I, II, and III (derivatives, integrals, vectors fields, line and multiple integrals, simple ordinary differential equations, and matrices) or Matrix and Linear Algebra. At UCF, completing these courses marks the transition from sophomore to junior standing. The course is often taken by seniors as well.

We follow the same approach of that course in this book and assume the reader knows no more than fundamental physics and mathematics at the junior level. The book attempts to be as self-contained as possible and starts with a concise description of the historic and conceptual developments that led to quantum mechanics, from Planck's black-body radiation to the Copenhagen interpretation. It then introduces vector spaces and operators utilizing Dirac's notation to familiarize the reader with the mathematical formulation used in quantum mechanics. The last foundational step is a chapter with a formal presentation of quantum mechanics postulates.

Chapters 5 to 9 cover key elements of quantum information with increasing level of complexity. After introducing qubits, gates and circuits, Bloch sphere representation, and other basic concepts and tools, the book moves on to key results such as the non-cloning theorem, teleportation, and quantum algorithms, from simple ones such as Deutsch's to more complicated ones such as Shor's and Grover's.

Those materials are the bulk of the course and the remaining chapters in the book are not always covered. The choice of which chapters to cover depends on the particular student cohort. For instance, when the majority of the students in the class is from computer science and engineering majors, chapters 10, 14, and 15 may be skipped. When the majority is from physics and optics & photonics, chapter 11 can be skipped and chapter 15 replaced by assigned term papers requiring a more detailed description of various physical realizations of qubits. It is also possible to skip certain sections in some chapters without compromising overall understanding. The book is organized in such a way that instructors adopting it in their courses may freely pick and choose what to cover according to the background and interests of their students. In the author's own experience, it is actually quite difficult to cover all sections of all chapters in a 15-week term.

All chapters but 12, 14, and 15 contain exercises to help students practice applying concepts and methods. These exercises were assigned as homework or featured in past mid-term and final exams. Solutions are provided only to instructors upon request.

At several points in the course, Qiskit, a python library for simulating and programming quantum computers, is used for practicing with quantum gates, quantum circuits, and for the graphical representation of Bloch spheres, circuits, quantum states and measurement outcomes. The book describes these activities; template codes are also available upon request. Python is such an accessible language that even students with very little familiarity with coding have been able to use the templates and complete the activities. An appendix explaining how to install and use Qiskit is included for convenience, but the reader should be aware that this package is constantly being updated, so the official online documentation should be always checked first.

Appendices are also included to help students with some linear algebra concepts and mathematical methods that are employed in the main text or are required for solving exercises. An appendix on the basics of computational complexity theory is also provided.

This book does not attempt to be a comprehensive treatise of quantum information processing, which is now a vast and yet still rapidly evolving subject. Like in the course that inspired it, the goal of this book is to provide just enough information to prepare undergraduates for more specialized, higher-level courses, and for research and employment opportunities that require basic knowledge of the subject. The focus is primarily on quantum computation because it covers foundational material for both quantum communications and quantum sensing. Therefore, the latter two areas of quantum information processing are only sparsely touched in this first edition and will hopefully be enhanced in the future.

References are provided at the end of all chapters for readers who want to gain a deeper understanding or learn topics in more details.

The author is in debt to more than one hundred students who have endured tough homework assignments and tests and yet found energy to provide suggestions and corrections to lecture notes and exercises. He would also like to acknowledge helpful suggestions from Drs. Hebin Li, Andrei Ruckenstein, and Bahaa Saleh.

E. R. Mucciolo

Winter Park, Florida
February, 2025

# About the Author

Eduardo R. Mucciolo is a professor of physics at the University of Central Florida (UCF). His current research falls at the interface between physics and computer science and often branches out into cryptography. He is a native of Brazil, where he received a B.S. and an M.S. in physics from the University of São Paulo. He received his Ph.D. in physics from MIT and was a postdoctoral fellow at NORDITA in Denmark. Before joining the faculty at UCF, he was an assistant and an associate professor at PUC-Rio in Brazil and held a visiting associate professor position at Duke University. He has over a hundred peer-reviewed publications and has given a similar number of invited talks. His research has been funded by FAPESP, CAPES, FAPERJ, and CNPq in Brazil, and by the NSF, the DOD, and the DOE in the USA. He is a fellow of the American Physical Society.

# CHAPTER 1

# Introduction

*What is Quantum Information Processing (QIP)?*

It is a collection of concepts, methods, and techniques that aim to process information (store, compute, transmit) by exploring two fundamental properties of quantum mechanical systems: interference and entanglement.

It is said that quantum mechanics has led to two revolutions so far.

- The first revolution started at around 1900 with the advent of foundational theories by Planck, Einstein, Bohr, Heisenberg, Dirac, and many others, leading to many scientific breakthroughs in atomic and molecular physics, solid-state physics, nuclear and particle physics, as well as technological advances such as lasers, transistors, magnetic storage, etc. Despite their tremendous impact on society, these new technologies did not exhaust the full potential of quantum mechanics.

- The second revolutions started in the late 1980s to mid 1990s by Bennett, Feynman, Deutsch, Shor, Grover, and others who showed that we can harness the power of quantum mechanics to solve problems of practical relevance that were considered too hard to tackle efficiently until then. The second revolution is about applying the rules and concepts learned during the first revolution to process information faster, more securely, and more accurately.

*What does it take to understand QIP?*

As much as mathematics is the language of physics, linear algebra is the language of QIP. Concepts are expressed in terms of "vectors", "scalers", "operators", "matrices", "tensors", "spaces", "transformations", etc. Fortunately, pioneers and founders of quantum mechanics have developed a nice and intuitive formulation that naturally infuses linear algebra. We will soon discuss this formulation.

Some knowledge of fundamental quantum phenomena is also important and we will tackle that in the next chapter.

Finally, the critical question: *what is QIP good for?*

We can break down QIP into three main areas, as far as applications are concerned:

1. quantum computing

2. quantum communication

3. quantum sensing.

In this order, they also reflect the level of "quantum power" needed, from high to low, as well as the level of difficulty in implementing them. These three areas share some common aspects such as the need to store information faithfully (quantum memories), to precisely control quantum states, and the inexorable susceptibility of quantum information to decoherence (the nemesis of QIP!).

Practical uses of quantum computing include: database search, cracking of cryptosystems, simulation of complex chemical systems (e.g., for drug development), solving linear algebra problems, and possibly artificial intelligence acceleration. Some of these applications require large and very low-noise quantum processors to work and are therefore somewhat futuristic. Hence, there is an intense search for other applications where one can take full advantage of the current generation of quantum processors, which are adequately named noisy intermediate-scale quantum (NISQ) devices.

Applications in quantum communications are already in use (including commercially) and boil down to eavesdropping-proof schemes of encryption key exchange. So far they have only been implemented from point A to point B; there are ongoing efforts to build networks relying on quantum exchange of information. This area is evolving rapidly.

Quantum sensing is mostly about enhancing the resolution and detection capabilities of scanners for medical, astronomical, and defense applications, among other uses. One one hand, some quantum sensing technologies are readily accessible and are making their way into commercial products; on the other hand, some applications require exquisite control or the use of very refined materials and are yet to be implemented in practice.

In this book, we mainly cover topics in quantum computation and quantum communications, from the fundamentals to applications. We try to balance depth and breath for both topics. Quantum sensing will be covered only superficially.

The reader should be aware that many other acronyms related to QIP are currently in use. In general, they have a broader scope than QIP and are more appealing to a non-technical audience, such as QISE (quantum information science and engineering) or QIST (quantum information science and technology). But, concepts and methods from QIP are at the core of these other denominations of the field.

# CHAPTER 2

# Basics of Quantum Phenomena

In this chapter we review the ideas and discoveries that lead to the development of quantum mechanics. Like other theories in physics, it was motivated by a series of experimental results that were at odds with the prevailing theories of the time. New concepts were introduced to explain those experiments, which in turn led to new experiments and more discoveries. Quantum mechanics took its current shape about 100 years ago and continues to provide a formidable framework for describing not only natural phenomena but also for leading us to new technologies, such as quantum information processing.

## 2.1  BLACKBODY RADIATION

Quantum mechanics was born in 1900 when Max Planck introduced the concept of quantization to explain why the radiation emitted by a "blackbody" [1] did not grow in intensity unboundedly as a function of the radiation frequency (the so-called ultraviolet catastrophe effect). This was a big deal because the prevailing physics theories at that time (Newton's mechanics, Maxwell's electromagnetism, and Boltzmann's statistical mechanics) predicted such a behavior, as shown in Fig. 2.1.

However, experiments results followed instead the behavior shown in Fig. 2.2.

---

[1] A blackbody is any macroscopic object in thermal equilibrium at a certain temperature. For an elementary description of blackbody radiation, check the OpenStax webpage `https://cnx.org/contents/NP3Ov7lW@2.49:0joA1o37@7/6-1-Blackbody-Radiation`.

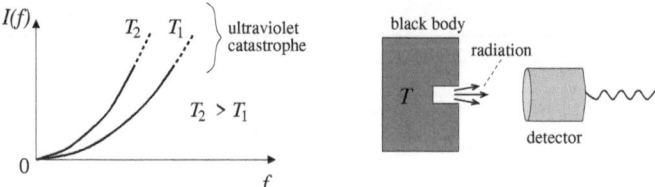

Figure 2.1 Left panel: intensity versus frequency for the radiation emitted by a blackbody for two distinct temperatures as predicted by classical physics. Right panel: a schematic illustration of a blackbody radiation detection setup. The detector is connected to a spectrum analyzer that produces the plot shown on the left panel.

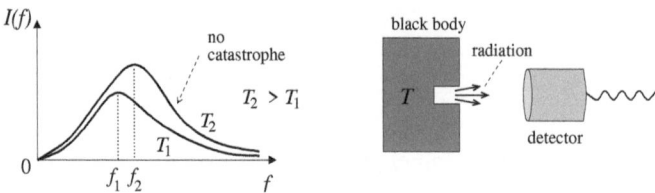

Figure 2.2 Intensity versus frequency for the radiation emitted by a blackbody observed in experiments.

The theory based on classical physics predicted that the intensity of the radiation should grow as a function of its frequency, while experiments showed that the intensity had a peak at a certain frequency, but then decayed for incresing frequencies. The experiments also showed that the higher the temperature, the higher the frequency where the peak is located.

Planck understood that the reason for $I(f)$ to grow unbounded with frequency in the classical theories had to do with the assumption that oscillations of the atoms in the blackbody had the same average energy, independently of their frequency. This assumption followed from Boltzmann's statistical mechanics, where the average energy was shown to be proportional to the temperature of the black body and independent of anything else.

Planck relaxed that assumption and made the oscillation energies to vary discretely and be proportional to their frequency,

$$E = nhf,$$

where $n = 0, 1, 2, 3, \ldots$ and $h$ is a constant (now named after Planck,

$6.62607 \times 10^{-34}$ m² kg s⁻¹). Based on this assumption, he derived an expression for the energy density of the blackbody radiation at temperature $T$,

$$u(f,T) = \frac{8\pi f^3}{c^2} \frac{1}{e^{hf/k_B T} - 1},$$

where $c$ is the speed of light in vacuum $(2.99792 \times 10^8$ m/s) and $k_B$ is the Boltzmann constant $(1.38065 \times 10^{-23}$ m²·kg·s²/K). His expression matched precisely the experimental data available! Moreover, it correctly described other features seen in experiments at that time, such as the dependence of the peak's frequency on temperature, as well as the so-called Stefan's total power law,

$$P \propto T^4,$$

where $P$ is the total power (i.e., energy per unit of time) of the radiation emitted by the blackbody.

The concept of quantized energy was as much revolutionary in 1900 as it was incomprehensible. Planck would win the 1918 Nobel Prize in Physics for his theory.

## 2.2 PHOTOELECTRIC EFFECT

However, the concept of quantization caught on. In 1905, Albert Einstein used it to put forward a theory of the photoelectric effect , which is the ejection of electrons from a metal surface when struck by light with frequencies above a threshold value (say, $f > f_0$), as shown in Fig. 2.3.

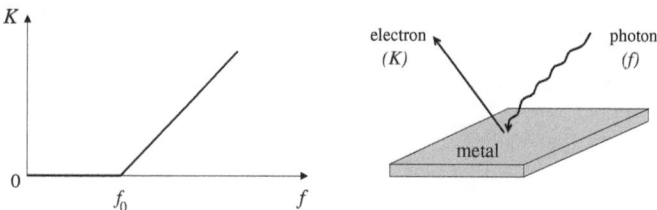

Figure 2.3  Left panel: photoelectron kinetic energy versus light frequency. Right panel: illustration of the photoelectric effect. The ejected electrons can be collected by a detector and their kinetic energy measured as a function of the incoming light's frequency, producing the plot shown on the left panel.

In Fig. 2.3, $K$ represents the electron's kinetic energy. Until Einstein's theory, it was unclear why such a minimum threshold frequency was needed and why is independent of the light intensity. Moreover, the experiments showed that $K$ depended only on the frequency of the light, and not on the light intensity either. More intense light did not produce faster electrons but light with higher frequency did.

Einstein explained this behavior by postulating that light itself is quantized, being formed by packets of energy (now called photons)

$$E = hf,$$

where $h$ is the same constant used by Planck . Then, by energy conservation, the ejected electron's kinetic energy obeys the relation

$$K = \begin{cases} hf - W = h(f - f_0) & \text{for} f > f_0 \\ 0 & \text{for} f \leq f_0 \end{cases},$$

where $W = hf_0$ is the so-called work function of the metal surface. The work function is a property of the metal, so different materials will have different work functions . It is the minimum energy required to free up an electron from the metal surface. Einstein won a Nobel Prize in Physics in 1921 for this explanation.

## 2.3 HYDROGEN ATOM

In 1913, Niels Bohr took the concept of quantization one step further. Starting from Rutherford's model of an atom (a massive positively charged nucleus surrounded by lightweight negatively charged electrons, like planets orbiting a star), Bohr quantized the orbital motion of the electron in the hydrogen atom[2] by assuming that its angular momentum can only vary discretely,

$$L = n\frac{h}{2\pi} = n\hbar, \tag{2.1}$$

where $n = 1, 2, 3, \ldots$. From this assumption and using elementary classical mechanics, it follows immediately that the electron can only take up discrete total energy values, $E_1$, $E_2$, $E_3$, etc. Bohr postulated that the transitions between quantized energy values in the hydrogen atom involved photons (to be emitted or absorbed) with well-defined, discrete energies:

$$hf = E_n - E_{n'}.$$

---

[2]Hydrogen was known to have only one electron orbiting around the nucleus.

Bohr used his theory to derive expressions for the spectral lines emitted by atomic hydrogen in the gas phase (see Fig. 2.4). His results matched nearly perfectly the experimental data available at the time, solving a long-standing puzzle in physics: Maxwell's equations, when combined with Newton's mechanics, predicted that electrons moving around the nucleus should emit radiation until they run out of kinetic energy, then falling into the nucleus. Thus Maxwell's and Newton's theories combined predicted that atoms were unstable! In 1922, Bohr received a Nobel Prize in Physics for his theory.

Figure 2.4  The visible spectral lines emitted by excited hydrogen atoms in gas state. The lines correspond to the Balmer series, where $n' = 2$). The wavelength increases from left to right. The four most intense lines have wavelengths 656 nm, 486 nm, 434 nm, and 410 nm, corresponding to $n = 3, 4, 5$, and 6, respectively. Image taken from LibreTexts Chemistry.

## 2.4  PARTICLE-WAVE DUALITY

An important aspect that underlies these early theories of quantization is that they apply to matter and radiation alike. The notion that electromagnetic radiation (light) can behave like a particle – the photon – was further solidified by experiments carried out in the early 1920s by Arthur Compton. He measured the dispersion of X-rays scattered off by electrons and showed that the wavelength of the X-rays increased after being scattered, something that only made sense if the X-ray beam behaved like a collection of particles with well-defined energy and momentum. This led Louis de Broglie in 1923 to propose the particle-wave duality relation:

$$p = \frac{hf}{c} = \frac{h}{\lambda}, \tag{2.2}$$

where $p$ is the linear momentum (a particle-like property), $f$ is frequency, and $\lambda$ is a wavelength (a wave-like property). Moreover, de Broglie proposed that such a relation holds for both matter (i.e., particles) and radiation (i.e., light).

For instance, Bohr's orbital quantization condition can be under-stood as a requirement that the electron's "wave" be stationary around the nucleus: since $L = mvr$ and $p = mv$, where $v$ is the electron velocity and $r$ is the radius of the orbit, using Eqs. (2.1) and (2.2), we can write

$$mvr = \frac{nh}{2\pi}$$

and

$$mv = \frac{h}{\lambda}.$$

Substituting the second equation into the first we obtain

$$n\lambda = 2\pi r.$$

The circumference of the electron's orbit must be a multiple of its wave-length, otherwise the orbit does not close onto itself and stationary motion cannot occur (i.e., it is unstable). See Fig. 2.5. Quantization is paramount to stability. Compton and de Broglie were also recipients of the Nobel Prize in Physics, in 1927 and 1929, respectively.

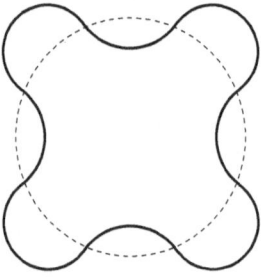

**Figure 2.5** A de Broglie standing wave whose wavelength is a quarter of the electron orbit's circumference ($n = 4$).

## 2.5 DOUBLE-SLIT EXPERIMENT

Is there another way to confirm the particle-wave duality more directly? The answer is yes! It is the double-slit experiment.

Suppose that we can prepare a beam of identical particles with a well-defined linear momentum (and thus a well-defined wavelength, per de Broglie's duality principle). Let us direct the beam toward a wall with two narrow slits close to each other and then use a screen on the other

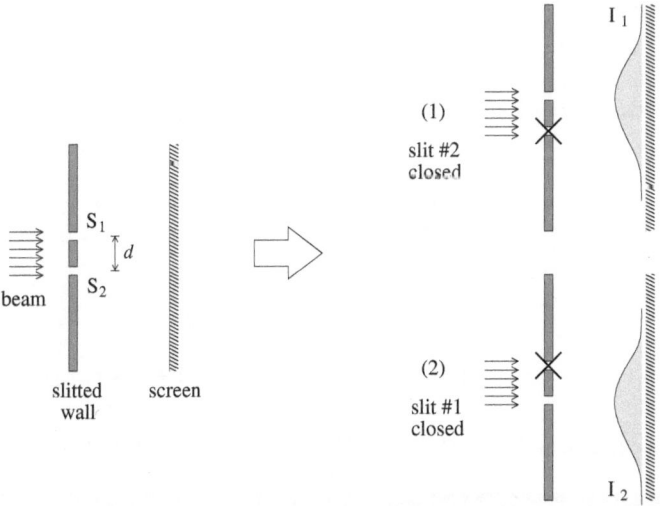

Figure 2.6 Left panel: the double-slit experimental setup. Right panel: outcome of the experiment for the cases when one of the slits is closed.

side of the wall to detect the particles that make it through the slits. See Fig. 2.6.

Firstly, let us consider two cases: (1) when slit #2 is closed and #1 is open; (2) when slit #1 is closed and #2 is open. On the screen, we will see a continuous and broad distribution of particle "hits", $I_1$ and $I_2$, respectively, each one peaked at a point in front of the corresponding open slit. We can understand the continuous distribution as due to particles being deflected as they pass through the open slit.

Secondly, consider a third case when both slits are simultaneously open. What will happen? There are two possible scenarios: the particle-like scenario, and the wave-like scenario, see Fig. 2.7.

In the particle-like scenario of Fig. 2.7a, we see a continuous and smooth distribution on the screen, centered at a point in front of the middle point between the slits. The total intensity is the sum of the intensities when the slits are individually closed. In the wave-like scenario of Fig. 2.7b, the distribution has peaks and troughs, with the peak amplitudes bounded by a continuous envelope, with the most intense peak at the point in front of the slits. We call this distribution an interference pattern. See Fig. 2.8 for an actual photograph of an interference pattern.

As long as the wavelength $\lambda$ of the beam and the distance $d$ between slits are of the same order of magnitude (and slits are thin enough), the

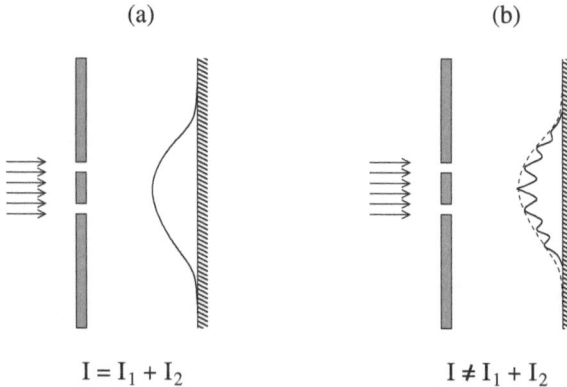

$$I = I_1 + I_2 \qquad\qquad I \neq I_1 + I_2$$

Figure 2.7 (a) Classical prediction. (b) Actual experimental result.

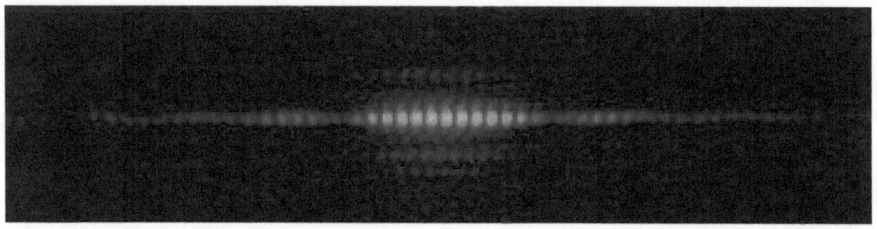

Figure 2.8 Image of the light emanating from a double slit after projection onto a screen. Image from Wikipedia Commons.

wave-like scenario (b) always prevails in experiments, no matter whether the beam consists of particles (e.g., electrons) or light (i.e., photons).

This experiment shows that interference is a key property of all quantum systems. As long as the beam wavelength and the slit separation are of the same order, the beam will split and then refocus, but in a way that reveals interference between the two secondary beams emitted by the slits. Superposition and interference are key fundamental properties that are heavily utilized in QIP. We will study their consequences in detail throughout this book.

## 2.6 THE BIRTH OF QUANTUM MECHANICS

After all these exciting developments in the first couple decades of the 20th century, physicists began to put together a more rigorous mathematical theory to describe and unify quantization, interference, and other phenomena that had been observed experimentally. They also began to build a framework to interpret the results predicted by this new theory. These were not trivial tasks. Consider that, in Newtonian mechanics, a particle can have both position and momentum simultaneously well defined. The situation is different for waves (e.g., electromagnetic waves), which can never be fully localized. How can these two apparently incompatible descriptions live side-by-side?

In 1926, Erwin Schrödinger published four papers with a mathematically complete description of the so-called wave mechanics of matter (nowadays known as the wave-based formulation of quantum mechanics). His theory provided a quantitatively accurate description of the electronic structure of the hydrogen atom and introduced the concept of wave functions. A year earlier, Werner Heisenberg had published a paper proposing a matrix formulation of quantum mechanics (which is now very much used in QIP), although very few people understood it at that time. He coined his theory "matrix mechanics" and introduced a wealth of concepts and tools. Schrödinger eventually showed that these two approaches, despite their mathematical differences, were equivalent.

A few years later, in 1928, Paul Dirac extended these theories to the relativistic domain, which is necessary when particles move close to the speed of light. And few more years later, he also introduced a very useful notation for quantum mechanics that we use to this day.

All three were awarded Nobel Prizes: Heisenberg in 1932 (although he only received in it 1933) and Schrödinger and Dirac in 1933.

Many other people made significant contributions in these early days of quantum mechanics, such as Wolfgang Pauli, Pascual Jordan, Enrico Fermi, Max Born, etc. One aspect that took a while to settle was the physical interpretation of the new theory. Eventually, the so-called Copenhagen interpretation (driven mainly by Niels Bohr and collaborators) prevailed and became widely accepted, although other interpretations still exist. Despite their differences, they mostly agree as far as the formal and practical mathematical aspects of quantum mechanics go. This is what matters most for a beginner student of QIP and therefore we will strictly follow the Copenhagen interpretation in this book.

## 2.7 AMPLITUDES AND PROBABILITIES

Let us go back to the double-slit experiment. What happens if instead of a dense beam of atomic particles we sent one particle at a time? To analyze this situation, we define the following probabilities:

$P(x)$ = probability of the particle hitting the screen at point of coordinate $x$ when both slits are open.

$P_1(x)$ = same, but when only slit $S_1$ is open.

$P_2(x)$ = same, but when only slit $S_2$ is open.

Our classical (non quantum) intuition tells us that

$$P(x) = P_1(x) + P_2(x).$$

But that is not what is seen in the experiments. Instead, the result is

$$P(x) = P_1(x) + P_2(x) + \Delta P_{12}(x),$$

where $\Delta P_{12}(x)$ represents interference and causes the peaks and troughs we discussed in the previous section (i.e., a sequence of intensity maxima and minima). How can we account for that?

Instead of composing probabilities, we need to use *amplitudes* . Let us introduce $a$ as the amplitude of a certain possible outcome of the experiment, where $a \in C$ (meaning, $a$ can be a complex number, with real and imaginary parts). It is convenient (but not required) that $|a| \leq 1$. How do we employ amplitudes? They are certainly not as intuitive as probabilities, but it turns out that they work similarly. We will now provide a non-rigorous explanation, but will get back to it in the next chapter after introducing a bit more math.

Suppose that the particle can take two distinct states or paths when passing through the slits toward the point of coordinate $x$ on the screen (i.e., passing through either slit $S_1$ or slit $S_2$, when they are both open). To each one of these distinct states we associate an amplitude, say, $a_1(x)$ and $a_2(x)$. Then, the total amplitude of the system can be written as the sum of these amplitudes, namely,

$$a(x) = a_1(x) + a_2(x).$$

How can we understand that? Notice that each individual amplitudes $a_1$ and $a_2$ corresponds to the outcome of a measurement that could be performed on the system to find out which path was taken. We expect only one outcome (either $S_1$ or $S_2$) in such a measurement.

We associate the square of the absolute value of each individual amplitude to the probability of observing that particular outcome (i.e., the particle passing through slit $S_1$ or slit $S_2$ and hitting a certain point on the screen) when measuring the system without any previous knowledge:

- $P_1(x) = |a_1(x)|^2$ = probability of the particle having passed through slit $S_1$ and hitting point $x$.

- $P_2(x) = |a_2(x)|^2 =$ probability of the particle having passed through slit $S_2$ and hitting point $x$.

When the system is free to choose between slits $S_1$ or $S_2$, we sum the amplitudes (the particle takes both paths at the same time!). The probability to hit point $x$ on the screen is now equal to

$$
\begin{aligned}
P(x) &= |a(x)|^2 = |a_1(x) + a_2(x)|^2 \\
&= |a_1(x)|^2 + |a_2(x)|^2 + a_1(x)a_2^*(x) + a_1^*(x)a_2(x).
\end{aligned}
$$

The extra terms are responsible for the interference effect,

$$
\Delta P_{12}(x) = a_1(x)a_2^*(x) + a_1^*(x)a_2(x).
$$

Notice that $\Delta P_{12}$ is always real, but it can be positive or negative. For instance, $a_1 = 1/3$ and $a_2 = -1/4$ results in $\Delta P_{12} = -1/6$ and $P < P_1 + P_2$ in this case.

The underlying concept in this construction is *superposition* , i.e., the ability of a quantum system to be simultaneously in distinct configurations. Namely, if there are $N$ possible outcomes for a certain measurement $X$, a quantum system can be on a state such that its amplitude is a sum of the amplitudes associated to each possible outcome, $a = a_1 + a_2 + \ldots + a_N$.

But there is a catch: the system may be in a superposition state, but if we try to measure $X$, we will only obtain one of the $N$ possible outcomes! It is only after we prepare the system multiple times in the same superposition state, and each time we perform the same measurement, that we start to see that some outcomes happen more often than others: outcome $k$ will happen with a frequency proportional to $P_k = |a_k|^2$, where $k = 1, 2, \ldots, N$.

There is a lot to digest here. To help fix these new concepts in our minds, imagine that the quantum system is an unbiased coin. Assume that $P_{\text{head}} = P_{\text{tail}} - 1/2$. We can set $a_{\text{head}} = a_{\text{tail}} = 1/\sqrt{2}$ for the sake of argument. In principle, it is possible to prepare this quantum coin in a state that is a superposition of head and tail. Yet, if we try to measure which face is up, we will either find head or tail, and not something in-between or both head and tail at the same time. To reveal the superposition state, we must measure a different property of the coin, one that is not directly related to which side is facing up. But to do that we need first to develop an appropriate mathematical formulation. We will do so by taking the electron as our quantum coin and focus on its spin, which is a property related to the electron's intrinsic magnetic moment, see Fig. 2.9 (every electron is a small – the smallest – magnet):[3]

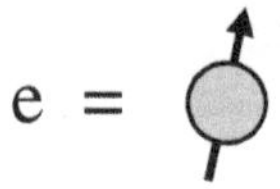

$$e = \enspace \phi$$

**Figure 2.9** Every electron carries a magnetic dipole moment.

In turns out that the electron spin, when measured, yields only two possible outcomes, regardless to how the measurement is performed. But being a vector in three spatial dimensions, the electron spin offers more possibilities of manipulation than a coin, allowing for more interesting concepts and phenomena, as we will soon find out.

## 2.8 REFERENCES AND FURTHER READING

1. Eisberg, R. and R. Resnick. 1985. *Quantum Physics of Atoms, Molecules, Solids, Nuclei, and Particles*, 2nd edition. John Wiley & Sons. Chapters 1-4.

2. Tipler, P. A. and Llewellyn, R. 2002. *Modern Physics*, 4th edition. W. H. Freeman. Chapters 3, 4, and 5.

3. Liboff, R. L. 2003. *Introductory Quantum Mechanics*, 4th edition. Addison Wesley. Chapter 2.

## 2.9 EXERCISES AND PROBLEMS

Useful quantities for calculations you may have to do:
$h = 6.63 \times 10^{-34}$ J·s
$k_B = 1.38 \times 10^{-23}$ J/K
$c = 3.00 \times 10^8$ m/s
$1 \text{ eV} = 1.602 \times 10^{-19}$ J
$1 \text{ Å} = 10^{-10}$ m.

---

[3]Maxwell's equations of electromagnetism say that any spinning electric charge distribution generates a magnetic moment. The tricky aspect is that electrons have no dimensions! They are point-like objects, as far as we can see in experiments and therefore should not have an intrinsic magnetic moment, according to classical physics. Their magnetic moment is quantum mechanical and is fully included in Dirac's relativistic formulation of quantum mechanics.

1. Planck's spectral density for the radiation emitted by blackbody as a function of the frequency is given by[4]

$$u(f) = \frac{8\pi h f^3}{c^3} \frac{1}{e^{hf/k_B T} - 1}.$$

This expression can be used to measure the temperature of a hard-to-reach source of radiation, such as a star or hot furnace, when the temperature and spectral density of a reference source are known. Optical pyrometers are based on this principle (typically, the spectral density is sampled at several frequencies).

(a) Call $T_r$ the temperature of a reference (i.e., known) radiation source and $T_u$ the unknown temperature of another source whose spectral density can be measured. Find an expression relating the ratio of the spectral densities of these two sources to their temperatures. Call this expression $r(f)$. How does $r(f)$ behave when $T_u = T_r$?

(b) About 70 years ago, Dicke, Penzias, and Wilson used a similar technique to find out that the Earth is surrounded uniformly in all directions by electromagnetic radiation as if the universe itself were a blackbody, which they determined to be at a temperature of 3 K. Their finding gave strong support to the big-bang theory for the origin of the universe. Find the predominant wavelength of the radiation of a blackbody at 3 K (*Hint*: find at which frequency Planck's formula has a maximum and then convert it to wavelength). Does this wavelength give you a clue for the technique they used in their experimental investigation?

2. In photoelectric effect experiments, one observes that electrons are ejected with different velocities, even when the light is monochromatic (i.e., has a well-defined frequency). Provide a plausible explanation of this phenomenon using Einstein's theory.

3. Light with a wavelength of 2000 Å hits an aluminum surface. For aluminum, 4.2 eV are necessary to remove an electron.

---

[4]This expression gives the amount of energy emitted by the blackbody radiation per unit of frequency and unit of volume. It differs with respect to other forms of the radiation distribution, such as when the density of emitted energy is defined per unit of wavelength. The two forms are related by the differential relation $u(f)df = u(\lambda)d\lambda$, recalling that $f = c/\lambda$.

    (a) What is the kinetic energy of the fastest photoelectron (i.e., the electron ejected by the light)?

    (b) What is the kinetic energy of the slowest photoelectron?

    (c) What is the cutoff wavelength for aluminum?

    (d) If the light intensity is 2.0 W/m$^2$, what is the number of photons per unit of time and area that hit the aluminum surface? *Hint*: intensity is the total amount of energy per unit of time and unit of area; start by finding out how much energy a single photon carries.

4. Why we do not experience wave-like phenomena such as interference and diffraction in our daily lives? You may illustrate your answers by considering an object of mass 40 g moving at 1000 m/s. What is its de Broglie wavelength?

5. Starting from Bohr's quantization of the angular momentum ($L = n\hbar$) and assuming that the proton (mass $m_p$ and charge $e$) stays stationary while the electron (mass $m_e$, charge $-e$) moves in a circular orbit around it, use elementary Newton's mechanics and Coulomb's law to:

    (a) Find an expression for the quantized radius of the orbit, $r_n$, of an electron in the hydrogen atom.

    (b) Find an expression for the quantized total energy of the electron $E_n$ in the hydrogen atom.

    (c) Find an expression relating the energy of the emitted (or absorbed) photon when the electron changes its orbit from $n$ to $n'$.

*Hints*: start by relating the radius of the orbit to the electron mass and its velocity; then use the angular momentum quantization to find the quantization of the radius; finally, relate the total energy (kinetic plus potential) to the radius.

# CHAPTER 3

# The Language of Quantum Mechanics

In this chapter we go through the essential mathematical tools needed to develop a quantitative description of quantum systems. The content will seem a bit dry and abstract, but gaining a good grasp of the concepts and the math of vectors and operators will make understanding quantum information processing a lot easier. So, take your time when studying this chapter. It will pay off.

## 3.1  VECTOR SPACES

We associate to each quantum system an abstract entity called a state vector, which we denote as

$$|\psi\rangle.$$

The letter $\psi$ labels the state of the quantum system. The brackets $|$ and $\rangle$ are used to remind us that we are dealing with a vector and not a scalar.

Why do we need vectors to describe the state of a quantum system? They main practical reason is to allow for superpositions, namely, the possibility to prepare a quantum system in classically distinct configurations at the same time, much like you can write a position vector in three dimensions as a sum of three vectors or components, $\vec{r} = x\,\hat{i} + y\,\hat{j} + z\,\hat{k}$, each one along one of the three orthogonal directions ($\hat{i}$, $\hat{j}$, and $\hat{k}$).

Why use the bracket notation instead of adding a little arrow on top of $\psi$ or changing its font to bold? The reason is that quantum state vectors have some properties that ordinary vectors in three-dimensional spaces do not have, and we want to make sure that the mathematical notation makes those additional properties to stand out. For instance, $|\psi\rangle$ is what we call a "ket" version of the state vector; we can also have a "bra" version, which we denote as $\langle\psi|$. They are analogous to a complex number and its conjugate.[1]

$|\psi\rangle$ lives in a vector space which we denote by $\mathcal{H}$ and call a Hilbert space.[2] Vectors in this space obey the following properties.

1. Addition: let $|\Omega\rangle = |\psi\rangle + |\phi\rangle$. If $|\psi\rangle, |\phi\rangle \in \mathcal{H}$, then $|\Omega\rangle \in \mathcal{H}$ as well.

2. Scalar multiplication: let $|\phi\rangle = c\,|\psi\rangle$, with $c \in \mathcal{C}$. If $|\psi\rangle \in \mathcal{H}$, then $|\phi\rangle \in \mathcal{H}$ as well.

3. Multiplication is distributive: $c\,(|\psi\rangle + |\phi\rangle) = c\,|\psi\rangle + c\,|\phi\rangle$.

4. Addition is commutative: $|\psi\rangle + |\phi\rangle = |\phi\rangle + |\psi\rangle$.

5. Addition is associative: $|\psi\rangle + (|\phi\rangle + |\Omega\rangle) = (|\psi\rangle + |\phi\rangle) + |\Omega\rangle$.

6. A null vector exists: $0 + |\psi\rangle = |\psi\rangle$ for every $|\psi\rangle \in \mathcal{H}$.

7. An inverse vector exists: $(-|\psi\rangle) + |\psi\rangle = 0$ for every $|\psi\rangle \in \mathcal{H}$.

8. There exists an inner product (see below for a definition).

(Rules 1 to 7 also apply to bra vectors.)

Let us play a bit with these vectors. Call $|\!\uparrow\rangle$ the vector representing the electron spin when it is pointing "up", and $|\!\downarrow\rangle$ when it is pointing "down". Quantum mechanics tells us that it is possible to prepare the electron in a superposition state,

$$|\psi\rangle = a_\uparrow\,|\!\uparrow\rangle + a_\downarrow\,|\!\downarrow\rangle,$$

where $a_\uparrow$ and $a_\downarrow$ are the amplitudes associated to "up" and "down", respectively. A fully spin-up state has $a_\uparrow = 1$ and $a_\downarrow = 0$; a fully spin-down one has $a_\uparrow = 0$ and $a_\downarrow = 1$. $|\!\uparrow\rangle$ and $|\!\downarrow\rangle$ represent very distinct configurations. How can we quantify that? Let us introduce a scalar product (also known as inner product) , such that:[3]

$$\langle\psi|\phi\rangle = c. \tag{3.1}$$

---

[1]Technically, $|\psi\rangle$ and $\langle\psi|$ correspond to the same state. They just live in dual vector spaces. Similarly to $z = a + ib$ and $z^* = a - ib$, which are defined by a single real tuple $(a, b)$. $z$ and $z^*$ have the same information content but we need both to compute the magnitude: $|z| = \sqrt{z\,z^*}$.

[2]Rigorously, a vector space is a Hilbert space only when an inner product is also defined.

[3]Through the inner product, vectors from dual spaces meet to produce a scalar.

To compute the scalar product, we take two vectors, one in its "ket" version and the other in its "bra" version, and join them together to produce a scalar (i.e., a number, which can be complex). To get some intuition about what a scalar product means, we can momentarily recall ordinary three-dimensional vectors. For instance, consider the velocity vectors shown in Fig. 3.1 and two of their "dot" products:

$$\vec{v}_1 = \qquad \vec{v}_2 = \qquad \vec{v}_3 = \longrightarrow$$

Figure 3.1  Three velocity vectors.

$$\vec{v}_1 \cdot \vec{v}_2 = |\vec{v}_1||\vec{v}_2| \cos(\theta_{12})$$
$$\vec{v}_1 \cdot \vec{v}_3 = |\vec{v}_1||\vec{v}_3| \cos(\theta_{13}).$$

Since $\theta_{13} = \pi/2$, $\vec{v}_1$ and $\vec{v}_3$ are orthogonal and $\vec{v}_1 \cdot \vec{v}_3 = 0$; and since $\theta_{12} < \pi/2$, $\vec{v}_1 \cdot \vec{v}_2 \neq 0$.

Going back to quantum mechanics, when two state vectors $|\psi\rangle$ and $|\phi\rangle$ correspond to completely different and distinguishable configurations, we expect their scalar product to vanish,

$$\langle\psi|\phi\rangle = 0.$$

We say that the two state vectors are "orthogonal". When they correspond to identical configurations, we expect instead the scalar product to have its maximum possible value, as if the vectors were "parallel" to each other,

$$\langle\psi|\phi\rangle = \|\psi\|\|\phi\|,$$

where we define the vector magnitude or norm as

$$\|\psi\| \equiv \sqrt{\langle\psi|\psi\rangle}$$

($\langle\psi|\psi\rangle \geq 0$ by definition).

For example, going back to spin states, since "up" is clearly distinguishable from "down" in a measurement of the spin's orientation, we can safely assume

$$\langle\uparrow\,|\downarrow\rangle = 0.$$

To facilitate calculations, we go one step further and assume that

$$\langle \uparrow \mid \uparrow \rangle = 1 \qquad \text{and} \qquad \langle \downarrow \mid \downarrow \rangle = 1. \tag{3.2}$$

This way of dealing with vectors through bras and kets is called the Dirac notation (a.k.a. "braket" notation). It is very useful, as we will soon find out. For instance, if

$$|\psi\rangle = a_\uparrow |\uparrow\rangle + a_\downarrow |\downarrow\rangle, \tag{3.3}$$

then

$$\langle \psi | = a_\uparrow^* \langle \uparrow | + a_\downarrow^* \langle \downarrow |,$$

where the ∗ indicates complex conjugation. We say that $\langle \psi |$ is the adjoint of $|\psi\rangle$. We can add kets to kets,

$$|\psi\rangle + |\phi\rangle,$$

and we can add bras to bras,

$$\langle \psi | + \langle \phi |,$$

but we cannot add kets to bras,

$$|\psi\rangle + \langle \phi | = \text{nonsense!}$$

However, we can take products of bras and kets,

$$\langle \psi | \phi \rangle \longrightarrow \text{inner product}$$

$$|\psi\rangle\langle\phi| \longrightarrow \text{outer product (to be explained soon)}.$$

Dirac's notation looks a bit weird but is very powerful. Consider Eq. (3.3). Taking the inner product of both sides with respect to the bra $\langle \uparrow |$, we have

$$\langle \uparrow | \psi \rangle = a_\uparrow \langle \uparrow | \uparrow \rangle + a_\downarrow \langle \uparrow | \downarrow \rangle = a_\uparrow$$

because of the relations shown in (3.2). Therefore,

$$|\langle \uparrow | \psi \rangle|^2 = |a_\uparrow|^2 = P_\uparrow,$$

which is the probability of finding the spin "up" upon measuring its orientation when the electron spin is in the state $\psi$.[4] Similarly,

$$|\langle \downarrow | \psi \rangle|^2 = |a_\downarrow|^2 = P_\downarrow.$$

---

[4]Of course, this probabilistic interpretation only makes sense when $\langle \psi | \psi \rangle = 1$, in which case $0 \leq P_\uparrow \leq 1$, $0 \leq P_\downarrow \leq 1$, and $P_\uparrow + P_\downarrow = 1$. Otherwise, we need to rescale $|a_\uparrow|^2$ by $\langle \psi | \psi \rangle$ to obtain the probability: $P_\uparrow = |a_\uparrow|^2 / \langle \psi | \psi \rangle$.

Thus we can extract amplitudes and probabilities from a state vector by taking inner products with suitable vectors.

Because the electron spin can only have two possible measurement outcomes, we only need two reference orthogonal vectors to span the entire vector space and be able to represent *any* superposition state. The vectors $|\uparrow\rangle$ and $|\downarrow\rangle$ can be those reference vectors, although they are not the only possible choice. Yet, once we settle on a choice, by judiciously picking the amplitudes in front of the vectors [e.g., the coefficients $a_\uparrow$ and $a_\downarrow$ in Eq. (3.3)], we can in principle describe *any* arbitrary state $|\psi\rangle$. We say that $\{|\uparrow\rangle, |\downarrow\rangle\}$ is a complete set of vectors for the two-dimensional spin space. Such a set is called a complete basis. A complete basis is not unique, even for a two-dimensional vector space. To see this point, let us go back to vectors in ordinary three-dimensional space: the Cartesian set $\{(1, 0, 0), (0, 1, 0), (0, 0, 1)\}$ is a good basis for that space, but so is the set $\left\{\left(\frac{1}{\sqrt{2}}, \frac{1}{\sqrt{2}}, 0\right), \left(\frac{1}{\sqrt{2}}, \frac{-1}{\sqrt{2}}, 0\right), (0, 0, 1)\right\}$ where we rotated the "x" and "y" basis vectors by 45°. The same occurs for quantum state vector spaces, where one can always design different but just-as-good basis through rotations.

A complete basis provides a handy way to compute inner products, especially if the basis vectors in the set are orthonormal (i.e., have norm 1 and are all orthogonal to each other). For example, consider two state vectors decomposed in the up and down basis,

$$|\psi\rangle = a_\uparrow|\uparrow\rangle + a_\downarrow|\downarrow\rangle$$
$$|\phi\rangle = b_\uparrow|\uparrow\rangle + b_\downarrow|\downarrow\rangle.$$

Then,

$$
\begin{aligned}
\langle\phi|\psi\rangle &= \left(b_\uparrow^*\langle\uparrow| + b_\downarrow^*\langle\downarrow|\right)\left(a_\uparrow|\uparrow\rangle + a_\downarrow|\downarrow\rangle\right) \\
&= b_\uparrow^* a_\uparrow\langle\uparrow\,|\,\uparrow\rangle + b_\uparrow^* a_\downarrow\langle\uparrow\,|\,\downarrow\rangle + b_\downarrow^* a_\uparrow\langle\downarrow\,|\,\uparrow\rangle + b_\downarrow^* a_\downarrow\langle\downarrow\,|\,\downarrow\rangle \\
&= b_\uparrow^* a_\uparrow + b_\downarrow^* a_\downarrow.
\end{aligned}
$$

The decomposition of state vectors on the same orthonormal basis allows us to compute their inner product by just multiplying amplitudes and adding the products. This is similar to what we do when taking the scalar products of three-dimensional position vectors written in terms of Cartesian coordinates:

$$\vec{v}_1 \cdot \vec{v}_2 = v_{1,x}\, v_{2,x} + v_{1,y}\, v_{2,y} + v_{1,z}\, v_{2,z}.$$

## 3.2  OPERATORS

The Dirac notation has one more surprise for us: outer products . Let us introduce

$$\hat{A} = |\psi\rangle\langle\phi|.$$

Notice that we can "contract" $\hat{A}$ from both left and right sides if we use a bra or a ket, respectively. Namely,

$$\langle\Omega|\,\hat{A} = \langle\Omega|\psi\rangle\langle\phi| = \alpha\langle\phi|$$

and

$$\hat{A}\,|\Sigma\rangle = |\psi\rangle\langle\phi|\Sigma\rangle = \beta|\psi\rangle,$$

where $\alpha = \langle\Omega|\psi\rangle$ and $\beta = \langle\phi|\Sigma\rangle$ are scalars. Thus, $\hat{A}$ "acting" on a bra produces another bra (times a scalar); $\hat{A}$ "acting" on a ket produces another ket (times a scalar as well). $\hat{A}$ is essentially an "operator" that takes a vector into another vector,

$$\hat{A}\,|v_1\rangle = |v_2\rangle$$

and

$$\langle u_1|\,\hat{A} = \langle u_2|.$$

We can conclude that the outer product of a bra with a ket results in an operator that can act on any vector in $\mathcal{H}$ to produce another vector in the same space.

The concept of an operator acting on vectors can be extended to situations when the operator cannot be expressed as single outer product. Thus, generally, it is more appropriate to think of operators as entities that act on vectors to produce other vectors.

In quantum mechanics, we only deal with linear operators:

$$\hat{A}\left(|\psi\rangle + |\phi\rangle\right) = \hat{A}\,|\psi\rangle + \hat{A}\,|\phi\rangle$$

and

$$\hat{A}\left(c\,|\psi\rangle\right) = c\left(\hat{A}\,|\psi\rangle\right),$$

where $c$ is a scalar (and similarly for bras).[5]

Not all operators are born equal! To understand their differences, let us define the adjoint operation for scalars, vectors, and operators:

$$|\psi\rangle \longrightarrow \langle\psi|$$

---

[5]To distinguish operators from scalars, we use capital letters for the former and endow them with a hat.

$$c \longrightarrow c^*$$
$$\hat{A} \longrightarrow \hat{A}^\dagger.$$

$\hat{A}^\dagger$ is called the Hermitian adjoint of $\hat{A}$. When $\hat{A} = \hat{A}^\dagger$, the operator is called Hermitian. When $\hat{A}^\dagger \hat{A} = \hat{I}$, where $\hat{I}$ is the identity operator (i.e., $\hat{I}$ is a do-nothing operator), $\hat{A}^{-1} = \hat{A}^\dagger$ (i.e., the inverse is equal to the Hermitian adjoint), in this case, $\hat{A}$ is called a unitary operator. When $\hat{A} \cdot \hat{A} = \hat{A}^2 = \hat{A}$, the operator is called a projector. These are the three most important types of operators in quantum mechanics.

### 3.2.1 Operators as matrices

There are a few more important facts about operators.

1. We can "sandwich" an operator with a bra and a ket to produce a scalar,
   $$\langle \psi | \hat{A} | \phi \rangle = c,$$
   implying $c^* = \langle \phi | \hat{A}^\dagger | \psi \rangle$ (a very useful property). In some specific contexts, such a sandwich is called a "matrix element".

2. When we have a complete basis for the vector space, and know how an operator acts on the vectors of the basis, we can associate a matrix to the operator. This is an extremely useful property because it turns abstract operator manipulations into matrix algebra. Here is an example of matrix representation of an operator using the basis $\{|\uparrow\rangle, |\downarrow\rangle\}$:
   $$\langle \uparrow | \hat{A} | \uparrow \rangle \equiv A_{11}, \qquad \langle \uparrow | \hat{A} | \downarrow \rangle \equiv A_{12}$$
   $$\langle \downarrow | \hat{A} | \uparrow \rangle \equiv A_{21}, \qquad \langle \downarrow | \hat{A} | \downarrow \rangle \equiv A_{22},$$

   which we can cast into a matrix form as
   $$\hat{A} = \begin{pmatrix} A_{11} & A_{12} \\ A_{21} & A_{22} \end{pmatrix}.$$

3. We can use a matrix notation for vectors as well: let
   $$|\uparrow\rangle = \begin{pmatrix} 1 \\ 0 \end{pmatrix} \qquad \text{and} \qquad |\downarrow\rangle = \begin{pmatrix} 0 \\ 1 \end{pmatrix}$$

   implying
   $$|\psi\rangle = a_\uparrow |\uparrow\rangle + a_\downarrow |\downarrow\rangle = \begin{pmatrix} a_\uparrow \\ a_\downarrow \end{pmatrix},$$

and

$$\langle\uparrow| = \begin{pmatrix} 1 & 0 \end{pmatrix} \qquad \text{and} \qquad \langle\downarrow| = \begin{pmatrix} 0 & 1 \end{pmatrix}$$

implying

$$\langle\psi| = a_\uparrow^*\langle\uparrow| + a_\downarrow^*\langle\downarrow| = \begin{pmatrix} a_\uparrow^* & a_\downarrow^* \end{pmatrix}.$$

4. Inner and outer products can be represented with matrices as well. Let

$$|\psi\rangle = \begin{pmatrix} a_\uparrow \\ a_\downarrow \end{pmatrix} \qquad \text{and} \qquad \langle\phi| = \begin{pmatrix} b_\uparrow^* & b_\downarrow^* \end{pmatrix}.$$

Then

$$\langle\phi|\psi\rangle = \begin{pmatrix} b_\uparrow^* & b_\downarrow^* \end{pmatrix} \begin{pmatrix} a_\uparrow \\ a_\downarrow \end{pmatrix} = b_\uparrow^* a_\uparrow + b_\downarrow^* a_\downarrow,$$

which is a scalar, i.e., a $1 \times 1$ matrix, and

$$|\psi\rangle\langle\phi| = \begin{pmatrix} a_\uparrow \\ a_\downarrow \end{pmatrix} \begin{pmatrix} b_\uparrow^* & b_\downarrow^* \end{pmatrix} = \begin{pmatrix} a_\uparrow b_\uparrow^* & a_\uparrow b_\downarrow^* \\ a_\downarrow b_\uparrow^* & a_\downarrow b_\downarrow^* \end{pmatrix},$$

which is a $2 \times 2$ matrix.

5. The action of an operator on a vector can also be performed in a matrix representation:

$$\hat{A}|\psi\rangle \longrightarrow \begin{pmatrix} A_{11} & A_{12} \\ A_{21} & A_{22} \end{pmatrix} \begin{pmatrix} a_\uparrow \\ a_\downarrow \end{pmatrix} = \begin{pmatrix} A_{11}\,a_\uparrow + A_{12}\,a_\downarrow \\ A_{21}\,a_\uparrow + A_{22}\,a_\downarrow \end{pmatrix}.$$

6. We can sum and multiply operators acting on the same Hilbert space:

$$\hat{A} + \hat{B} = \hat{C}$$
$$\hat{A}\,\hat{B} = \hat{D}.$$

Using a matrix representation, we have

$$\begin{pmatrix} A_{11} & A_{12} \\ A_{21} & A_{22} \end{pmatrix} + \begin{pmatrix} B_{11} & B_{12} \\ B_{21} & B_{22} \end{pmatrix} = \begin{pmatrix} A_{11} + B_{11} & A_{12} + B_{12} \\ A_{21} + B_{21} & A_{22} + B_{22} \end{pmatrix}$$
$$= \begin{pmatrix} C_{11} & C_{12} \\ C_{21} & C_{22} \end{pmatrix}$$

and

$$
\begin{pmatrix} A_{11} & A_{12} \\ A_{21} & A_{22} \end{pmatrix} \begin{pmatrix} B_{11} & B_{12} \\ B_{21} & B_{22} \end{pmatrix}
$$

$$
= \begin{pmatrix} A_{11}B_{11} + A_{12}B_{21} & A_{11}B_{12} + A_{12}B_{22} \\ A_{21}B_{11} + A_{22}B_{21} & A_{21}B_{12} + A_{22}B_{22} \end{pmatrix}
$$

$$
= \begin{pmatrix} D_{11} & D_{12} \\ D_{21} & D_{22} \end{pmatrix}.
$$

However, it is of fundamental importance to know that operators do not necessarily commute under multiplication: in general $\hat{A}\hat{B} \neq \hat{B}\hat{A}$. When operators commute, something special happens, as we will see below.

## 3.3   SHORT BREAK TO TAKE A BREATH AND TO SUMMARIZE

What have we learned so far?

- The state of a quantum system can be represented by a vector in a Hilbert space.

- To each vector we associate a dual one (e.g., a ket to a bra and vice versa).

- State vectors can be decomposed in terms of basis vectors; the most useful bases are those comprising normalized, mutually orthogonal vectors.

- In a decomposition, the amplitude corresponding to a certain basis vector can be related to the probability amplitude of finding the system in the configuration associated to that basis vector.

- Using a vector and its dual, we can compute scalar (inner) products, as well as outer products (which function as operators).

- Using an orthonormal basis, we can represent vectors as single-column or single-row matrices and operators as square matrices.

## 3.4   MORE ABOUT OPERATORS

Now that we learned the basics, we can start using a more general formulation. For instance, we can denote an orthonormal basis of an

$N$-dimensional Hilbert space as

$$\{|\phi_k\rangle\}_{k=1,\ldots,N},$$

where[6]

$$\langle\phi_k|\phi_{k'}\rangle = \delta_{k,k'}.$$

Then, for any state $|\psi\rangle$ in that space we can write

$$|\psi\rangle = \sum_{k=1}^{N} a_k|\phi_k\rangle,$$

where $a_k = \langle\phi_k|\psi\rangle$ (check it!). Also, for any operator $\hat{A}$ acting on vectors in that space we can write,

$$\hat{A} = \sum_{k=1}^{N}\sum_{k'=1}^{N} A_{k,k'}|\phi_k\rangle\langle\phi_{k'}|, \tag{3.4}$$

where $A_{k,k'} = \langle\phi_k|\hat{A}|\phi_{k'}\rangle$ (check it as well!).

### 3.4.1 Trace of an operator

Because we associate matrices to operators, we can easily define the trace of an operator: Starting from Eq. (3.4) and the orthonormal basis $\{|\phi_k\rangle\}$, we write

$$\text{tr}[\hat{A}] \equiv \sum_{k=1}^{N}\langle\phi_k|\hat{A}|\phi_k\rangle = \sum_{k=1}^{N} A_{k,k}.$$

In words: the trace of an operator is the sum of its diagonal matrix elements. It turns out that the trace of an operator is independent of the basis! Namely, any basis decomposition yields the same trace. It is thus an intrinsic property of the operator.

### 3.4.2 Eigenvalues and eigenvectors of an operator

Another intrinsic property of an operator is its "spectrum", which is the collection of its eigenvalues. Let us define what an eigenvalue is.

Consider a vector $|v\rangle$ such that, for an operator $\hat{A}$,

$$\hat{A}|v\rangle = a|v\rangle,$$

---

[6]Here, $\delta_{k,k'}$ denotes the Kronecker delta : $\delta_{k,k'} = \begin{cases} 1 \text{ if } k = k' \\ 0 \text{ otherwise} \end{cases}$.

where $a$ is some scalar. In words, the action of the operator on the vector produces the same vector, up to a constant factor. When this happens, $|v\rangle$ is called an eigenvector of the operator $\hat{A}$ with a corresponding eigenvalue $a$. The set of all eigenvalues of an operator is called its spectrum. Operators acting on an $N$-dimensional Hilbert space can have as many as $N$ distinct eigenvalues and corresponding eigenvectors. However, it is not uncommon to have the same eigenvalue showing up more than once, which is a phenomenon called degeneracy. Eigenvalues in a degenerate subset are all equal but their corresponding eigenvectors are not.

Here are a couple important facts about eigenvalues.

- Hermitian operators have only real eigenvalues:

  if $\hat{A} = \hat{A}^\dagger$ and $\hat{A}|v\rangle = a|v\rangle$, then $a = a^*$.

- The eigenvalues of unitary operators are complex numbers of magnitude 1:

  if $\hat{A}^\dagger = \hat{A}^{-1}$ and $\hat{A}|v\rangle = a|v\rangle$, then $|a| = 1$.

There is a fundamental result from linear algebra that is extensively used in quantum mechanics. It is called the spectral theorem: let us define a linear operator as normal when $\hat{A}\hat{A}^\dagger = \hat{A}^\dagger\hat{A}$ (i.e., the operator commutes with its adjoint). Then, *for every normal operator acting on a Hilbert space of finite dimension, there is an orthonormal basis formed by the operator's eigenvectors.*

It turns out that Hermitian and unitary operators are always normal! Hence, an important application of the spectral theorem is that every Hermitian operator provides an orthonormal basis through its eigenvectors. This also means that every Hermitian operator can be "diagonalized", namely, be decomposed in its own eigenbasis: if $\hat{A}|v_k\rangle = a_k|v_k\rangle$, then

$$\hat{A} = \sum_{k=1}^{N} a_k |v_k\rangle\langle v_k|.$$

Notice that there are no cross terms in the decomposition (this is what we mean by "diagonal"). To understand how this translates into a matrix representation, we need one more result. One can show that Hermitian operators can be diagonalized by unitary operators: if $\hat{A} = \hat{A}^\dagger$, then there exists a $\hat{U}$ such that $\hat{A} = \hat{U} \cdot \hat{\Lambda} \cdot \hat{U}^\dagger$, where $\hat{U}^\dagger = \hat{U}^{-1}$ and $\hat{\Lambda}$ is a diagonal matrix containing the eigenvalues of $\hat{A}$. The columns of $\hat{U}$ contain the corresponding eigenvectors of $\hat{A}$.

As an example, consider the $3 \times 3$ Hermitian matrix

$$\hat{A} = \begin{pmatrix} -1 & 5 & 2 \\ 5 & -1 & 2 \\ 2 & 2 & 2 \end{pmatrix},$$

which is diagonalized by the unitary matrix

$$\hat{U} = \begin{pmatrix} 1/\sqrt{3} & 1/\sqrt{6} & -1/\sqrt{2} \\ 1/\sqrt{3} & 1/\sqrt{6} & 1/\sqrt{2} \\ 1/\sqrt{3} & -2/\sqrt{6} & 0 \end{pmatrix}$$

and yields the diagonal matrix

$$\hat{\Lambda} = \hat{U}^\dagger \cdot \hat{A} \cdot \hat{U} = \begin{pmatrix} -6 & 0 & 0 \\ 0 & 0 & 0 \\ 0 & 0 & 6 \end{pmatrix}.$$

Notice that

$$\hat{A} \begin{pmatrix} \frac{1}{\sqrt{3}} \\ \frac{1}{\sqrt{3}} \\ \frac{1}{\sqrt{3}} \end{pmatrix} = 6 \times \begin{pmatrix} \frac{1}{\sqrt{3}} \\ \frac{1}{\sqrt{3}} \\ \frac{1}{\sqrt{3}} \end{pmatrix},$$

$$\hat{A} \begin{pmatrix} \frac{1}{\sqrt{6}} \\ \frac{1}{\sqrt{6}} \\ \frac{-2}{\sqrt{6}} \end{pmatrix} = 0 \times \begin{pmatrix} \frac{1}{\sqrt{6}} \\ \frac{1}{\sqrt{6}} \\ \frac{-2}{\sqrt{6}} \end{pmatrix},$$

$$\hat{A} \begin{pmatrix} \frac{-1}{\sqrt{2}} \\ \frac{1}{\sqrt{2}} \\ 0 \end{pmatrix} = -6 \times \begin{pmatrix} \frac{-1}{\sqrt{2}} \\ \frac{1}{\sqrt{2}} \\ 0 \end{pmatrix}.$$

Interestingly, notice also that $\text{Tr}[\hat{A}] = \text{Tr}[\hat{\Lambda}] = 0$. In fact, as noted earlier for operators, the trace of a square matrix is an *invariant*, which means that it does not change upon unitary transformations of the matrix (i.e., upon a change of basis). The trace of a matrix is always the sum of its eigenvalues.

One final note in regard to eigenvalues and eigenvectors: when two operators commute, they share the same eigenvectors. However, their eigenvalue sets do not need to be equal. For instance, if $\hat{A}\hat{B} = \hat{B}\hat{A}$ and $\hat{A}|\phi_n\rangle = a_n|\phi_n\rangle$, then $\hat{B}|\phi_n\rangle = b_n|\phi_n\rangle$ but $b_n$ is not necessarily equal to $a_n$.

## 3.5 REFERENCES AND FURTHER READING

1. Zettili, N. 2009. *Quantum Mechanics, Concepts and Applications,* 2nd edition. John Wiley & Sons. Sections 2.1-2.5.

2. Shankar, R. 1994. *Principles of Quantum Mechanics,* 2nd edition. Plenum Press. Sections 1.1-1.9.

3. Schumacher B. and M. Westmoreland. 2010. *Quantum Processes, Systems, and Information.* Cambridge Univ. Press. Chapter 3.

4. Nielsen M. A. and I. L. Chuang. 2000. *Quantum Computation and Quantum Information.* Cambridge Univ. Press. Section 2.1.

## 3.6 EXERCISES AND PROBLEMS

1. Consider the three 3-dimensional vectors

$$\vec{a} = (1, 2, 0), \quad \vec{b} = (-2, 1, 0), \quad \vec{c} = (1, 1, 1).$$

Let the scalar (inner) product have the standard form

$$\vec{u} \cdot \vec{v} = u_1 v_1 + u_2 v_2 + u_3 v_3$$

for $\vec{u} = (u_1, u_2, u_3)$ and $\vec{v} = (v_1, v_2, v_3)$.

   (a) Normalize the vectors $\vec{a}$, $\vec{b}$, and $\vec{c}$ (namely, find $\vec{a}\,' = \vec{a}/|\vec{a}|$, etc.).

   (b) Verify that the set $\{\vec{a}\,', \vec{b}\,', \vec{c}\,'\}$ does not form an orthogonal basis.

   (c) Keeping $\vec{a}\,'$ fixed, find an orthonormal basis (i.e., find two other vectors $\vec{b}\,''$ and $\vec{c}\,''$ such that the triplet of vectors forms an orthonormal basis). *Hint:* look for the Gram-Schmidt process from linear algebra.

2. A linear vector space is a very general concept that can be applied to a variety of contexts. Consider the set of functions

$$\phi_{2n}(x) = \sin(nx) \quad \text{and} \quad \phi_{2n+1}(x) = \cos(nx),$$

where $n = 0, 1, 2, \ldots$ and $x \in [0 : 2\pi)$. It turns out that the set $\{\phi_m(x)\}_{m=0,1,2,\ldots}$ spans the entire linear vector space of continuous real functions with domain in the interval $[0 : 2\pi)$; namely, for such functions, it always possible to write

$$f(x) = \sum_{m=0}^{\infty} \alpha_m \phi_m(x) \qquad (3.5)$$

for some set of real coefficients $\{\alpha_m\}_{m=0,1,2,...}$. Let us associate to each function $\phi_m(x)$ a ket $|\phi_m\rangle$ and a bra $\langle\phi_m|$.

(a) Rewrite Eq. (3.5) as ket and bra equations, assuming that $|f\rangle$ and $\langle f|$ represent the function $f(x)$.

(b) What is the dimensionality of this vector space?

(c) Show that

$$\langle f|g\rangle \equiv \int_0^{2\pi} dx\, f(x) g(x)$$

is an acceptable scaler (inner) product of this vector space.
*Hint:* you need to prove that $\langle f|f\rangle \geq 0$ and $\langle f|(a|g\rangle + b|h\rangle) = a\langle f|g\rangle + b\langle f|h\rangle$ for any $|f\rangle$, $|g\rangle$, $|h\rangle$, $a$, and $b$.

(d) Is the basis $\{|\phi_m\rangle\}_{m=0,1,2,...}$ orthogonal? Is the basis orthonormal? If not, how do you turn it into an orthonormal one?

3. Consider all polynomials of degree 3 in a real variable $x$ confined to the interval $[0{:}1]$, $P_a(x) = a_0 + a_1 x + a_2 x^2 + a_3 x^3$, where $a = (a_0, a_1, a_2, a_3)$ are real numbers. Let the operation $P_a + P_b$ denote ordinary addition.

(a) Show that the set of all $P_a$ when combined with ordinary addition constitutes a linear vector space. In your proof, use the Dirac notation with a suitable set of variables or indices to label the vectors.

(b) Show that

$$\int_0^1 dx\, P_a(x) P_b(x)$$

works a scalar (inner) product for this vector space.

(c) What is the dimensionality of this vector space?

(d) What would constitute a good orthonormal basis for this vector space?

4. Show that Hermitian operators can only have real eigenvalues.

5. Show that unitary operators can only have eigenvalues with magnitude 1.

6. Show that if $\hat{A}$ commutes with $\hat{B}$, then they share an eigenbasis.

7. Let $\{|\uparrow\rangle |\downarrow\rangle\}$ constitute an orthonormal basis of the two-dimensional Hilbert space of a spin-1/2 electron. Consider a state

$$|\psi\rangle = a_\uparrow |\uparrow\rangle + a_\downarrow |\downarrow\rangle,$$

where $a_\uparrow$ and $a_\downarrow$ are the probability amplitudes. Show that if $\langle \psi | \psi \rangle = 1$, it follows that $|a_\uparrow|^2 + |a_\downarrow|^2 = 1$. How would you define the probabilities of measuring the spin in the states $|\uparrow\rangle$ and $|\downarrow\rangle$ if $\langle \psi | \psi \rangle \neq 1$?

8. Compute analytically the eigenvalues and eigenvectors of the matrix

$$\begin{pmatrix} 0 & 0 & 1 \\ 0 & 0 & 0 \\ 1 & 0 & 0 \end{pmatrix}.$$

Is this matrix Hermitian? Is it unitary? Is it a projector?

9. An important manifestation of "quantumness" is tunneling, namely, the ability of a quantum system to transit between two configurations separated by an energy barrier higher than the available energy to the system. Consider, for instance a single particle in a double well, as show in Fig. 3.2 below, where the total energy is assumed to be lower than the barrier height. Denote as $|L\rangle$ and $|R\rangle$ the state vectors associated to the particle being on the left and right wells, respectively. Assume that these vectors are orthonormal.

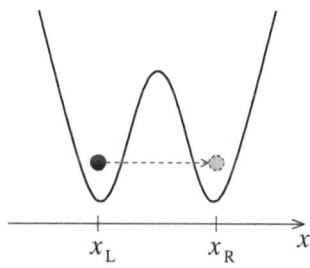

Figure 3.2  Double-well quantum system with a single particle.

(a) Show that the operator $\hat{X} = |L\rangle\langle R| + |R\rangle\langle L|$ is Hermitian.

(b) Find the eigenvectors (eigenkets) and eigenvalues of $\hat{X}$.

(c) Show that the operators $\hat{Y} = -i|L\rangle\langle R| + i|R\rangle\langle L|$ and $\hat{Z} = |L\rangle\langle L| - |R\rangle\langle R|$ are also Hermitian.

(d) Find the eigenvalues of $\hat{Y}$ and $\hat{Z}$.

(e) Show that $\hat{X}\hat{Y} = i\hat{Z}$, namely, that the sequential operation of $\hat{X}$ and $\hat{Y}$ on any vector is equivalent to operating with $i\hat{Z}$ on the same vector. What happens when the order of the product of $\hat{X}$ and $\hat{Y}$ is reversed?

(f) Show that $\hat{Z}\hat{X} = i\hat{Y}$ and $\hat{Y}\hat{Z} = i\hat{X}$.

(g) Show that $\hat{I} = |L\rangle\langle L| + |R\rangle\langle R|$ is an identity operator in the context of the double well system.

(h) Show that $\hat{P}_L = |L\rangle\langle L|$ and $\hat{P}_R = |R\rangle\langle R|$ are projection operators.

(i) What happens when you take integer powers of the operators $\hat{X}$, $\hat{Y}$, and $\hat{Z}$?

(j) Find matrix representations of the operators $\hat{P}_L$, $\hat{P}_R$, $\hat{I}$, $\hat{X}$, $\hat{Y}$, and $\hat{Z}$. What are the traces of these operators? *Hint*: use the eigenbasis $\{|L\rangle, |R\rangle\}$ to represent your matrices.

# CHAPTER 4

# Quantum Mechanics

Now that we went through the math and the notation, we can introduce the fundamentals of quantum mechanics in a formal and precise way.

## 4.1  POSTULATES OF QUANTUM MECHANICS

We will begin with five postulates that establish the theoretical foundations and mathematical framework of quantum mechanics.

I. The state of a quantum system is represented by a vector in a Hilbert space,

$$|\psi(t)\rangle.$$

*Note*: we added a time variable $t$ to indicate that we are considering the possibility of state vectors changing over time.

*Note*: contrast this representation with that in Newton's mechanics where the state of a classical system is specified by position and velocity coordinates.

II. Observable and measurable quantities are represented by Hermitian operators.

*Example*: if the system consists of a single particle in three spatial dimensions, its position and linear momentum will be represented by operators $\hat{x}$, $\hat{y}$, $\hat{z}$, $\hat{p}_x$, $\hat{p}_y$, and $\hat{p}_z$. (The hats help us distinguish scalars from operators.) Position and space coordinates are not variables in quantum mechanics but rather observables represented by operators.

III. If a system is in a state $|\psi\rangle$,[1] the measurement of a quantity represented by an operator $\hat{\Omega}$ will always yield one of its eigenvalues. For instance, the eigenvalue $\omega$ will occur with probability $P(\omega) = |\langle\omega|\psi\rangle|^2$, where $|\omega\rangle$ is the eigenvector of $\hat{\Omega}$ associated to $\omega$, namely,

$$\hat{\Omega}|\omega\rangle = \omega|\omega\rangle.$$

We say that the state of the system "collapses" to $|\omega\rangle$ after the measurement.

*Note*: this postulate is a profound departure from classical physics. See below for comments.

IV. The state vector $|\psi(t)\rangle$ obeys the Schrödinger equation

$$i\hbar\frac{d}{dt}|\psi(t)\rangle = \hat{H}|\psi(t)\rangle,$$

where $\hat{H}$ is the Hamiltonian operator of the system.

*Note*: this equation plays a similar role to Newton's second law in classical mechanics and sets the dynamics of the quantum system. It relates the interactions existing in the system (r.h.s. of the equation) to the change in its state vector (l.h.s. of the equation).

V. When two systems are combined as one system, the resulting Hilbert space is the tensor product of the Hilbert spaces of the subsystems: $\mathcal{H} = \mathcal{H}_1 \otimes \mathcal{H}_2$. If system #1 is in a state $|\psi_1\rangle$ and subsystem #2 is in a state $|\psi_2\rangle$, the total system is in a product state $|\psi_1\rangle \otimes |\psi_2\rangle$. Thus, if the Hilbert space of system #1 has dimension $d_1$ and its counterpart for system #2 has dimension $d_2$, the composed system has a Hilbert space of dimension $d = d_1 d_2$. The state of a system composed of subsystems does not need to always be a product (i.e., separable): entanglement can occur.

*Note*: this postulate can be straightforwardly generalized to any number of subsystems.

There is a lot to comment and explain!

## 4.1.1 Comments on postulate I

A vector space is an abstract concept but we can try to make physical sense of it by choosing an appropriate basis. The choice of such a basis

---

[1]We implicitly assume that $|\psi\rangle$ is normalized.

is guided by the physical characteristics of the particular system under consideration, and that is where the math meets the physical world.

*Examples*:

1. If we want to describe the spin of an electron or nucleon ($s = 1/2$), then we choose the basis vectors

   $|\uparrow\rangle$ for spin "up" (aligned)

   $|\downarrow\rangle$ for spin "down" (anti-aligned)

   in reference to some direction in the three-dimensional space. For this system, $\{|\uparrow\rangle, |\downarrow\rangle\}$ is a good orthonormal basis and, as seen before, the Hilbert space is two-dimensional.

2. If the system is a single-electron atom with orbital states $\phi_1$, $\phi_2$, $\phi_3$, ..., we pick those orbitals as basis, $\{|\psi_1\rangle, |\phi_2\rangle, |\phi_3\rangle, ...\}$, see Fig. 4.1. There is a discrete but infinite number of such orbital states (thus the Hilbert space of such a system has infinite dimension).

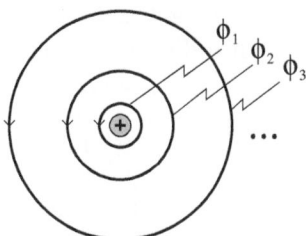

Figure 4.1  Schematic representation of orbital states in an atom.

Let $E_1$, $E_2$, $E_3$, ... be the energies associated to these (stationary) orbital states of the atom. When $E_1$ is near to $E_2$ but $E_3$, $E_4$, ... are much further out (i.e., higher in energy), we can neglect all but the states $|\phi_1\rangle$ and $|\phi_2\rangle$, effectively approximating the Hilbert space to two dimensional.

Inserting an overall phase on the state vector does not represent a change in the state of a system, namely, if

$$|\psi\rangle = e^{i\theta}|\phi\rangle,$$

with $\theta \in \mathcal{R}$, then $|\psi\rangle$ and $|\phi\rangle$ describe the same quantum state. The phase $\theta$ is not observable.

However, a relative phase can completely alter a state, namely, if

$$|\psi\rangle = |\uparrow\rangle + |\downarrow\rangle$$

and

$$|\phi\rangle = |\uparrow\rangle + e^{i\pi/4}|\downarrow\rangle,$$

then $|\psi\rangle \neq |\phi\rangle$ both mathematically and physically; in this case, the relative phase $\pi/4$ is actually observable (usually via an interference experiment).

It turns out that when a quantum system interacts with another one, especially with a larger one that we do not know or care how to describe in detail, a state vector is not enough to characterize the system's state because quantum information is lost to the other system. Later, we will describe a way to cope with such a situation.

### 4.1.2 Comments on postulate II

Operators do not necessarily commute. At first, this seems to be just a mathematical inconvenience, but it turns out to have a profound physical interpretation and some important practical implications. Let us define what we mean by non commutability more precisely: it is the dependence on order with which we apply operators on vectors. In general,

$$\hat{A}(\hat{B}|\phi\rangle) \neq \hat{B}(\hat{A}|\phi\rangle)$$

for two operators $\hat{A}$ and $\hat{B}$ and a state vector $|\phi\rangle$. When two operators do not commute, the uncertainties in their measurements are related to each and lower bounded. Heisenberg was the first one to notice this property, which is now called the uncertainty principle.

To understand this point, let us define the average (also known as expectation value) of a Hermitian operator $\hat{A}$ on a state $|\psi\rangle$ as[2]

$$\langle\hat{A}\rangle \equiv \langle\psi|\hat{A}|\psi\rangle.$$

The definition of variance of the same operator with respect to $|\psi\rangle$ is straightforward:

$$\begin{aligned}
\mathrm{var}(\hat{A}) &\equiv \langle\psi|(\hat{A} - \langle\hat{A}\rangle\hat{I})^2||\psi\rangle \\
&= \langle\psi|\hat{A}^2|\psi\rangle - \langle\hat{A}\rangle^2 \\
&= \langle\hat{A}^2\rangle - \langle\hat{A}\rangle^2,
\end{aligned}$$

---

[2]Notice the implicit dependence on the particular state used to compute the average.

where $\hat{I}$ is the identity operator. The uncertainty of the operator $\hat{A}$ with respect to the state $|\psi\rangle$ is defined as

$$\Delta A \equiv \sqrt{\mathrm{var}(\hat{A})}.$$

Let $\hat{A}$ and $\hat{B}$ be two Hermitian operators . Then, it is possible to prove that

$$(\Delta A)(\Delta B) \geq \frac{1}{2}|\langle\hat{\Gamma}\rangle|, \tag{4.1}$$

where $\hat{\Gamma} = \hat{A}\hat{B} - \hat{B}\hat{A} \equiv [\hat{A}, \hat{B}]$. The latter is called the commutator of $\hat{A}$ and $\hat{B}$ . In particular, when $\hat{A}$ and $\hat{B}$ are so-called canonically conjugated operators (a terminology borrowed from classical mechanics), namely, when $[\hat{A}, \hat{B}] = i\hbar\hat{I}$, then

$$(\Delta A)(\Delta B) \geq \frac{\hbar}{2}. \tag{4.2}$$

This is the case for the position and linear momentum operators, leading to the famous relation $\Delta x\,\Delta p \geq \hbar/2$ (Heisenberg's uncertainty principle). But what does the uncertainty principle mean?

Simply put, the uncertainty principle says that the more certainty we have about the value of $\hat{A}$, the less we know about the value of $\hat{B}$, and vice versa. By more certainty we mean decreasing values of $\Delta A$: when $\Delta A \to 0$, by the relation (4.2), we must have $\Delta B \to \infty$, in which case the fluctuations in $\hat{B}$ from one measurement to another are so large that we learn nothing about $\hat{B}$ while we are certain about $\hat{A}$.

From a physics standpoint, it means that the quantities associated to the operators $\hat{A}$ and $\hat{B}$ cannot be simultaneously determined with complete certainty. For instance, if we know that an electron is localized within a region of width $\Delta x$, we cannot expect to learn about its linear momentum with precision better than $\Delta p = \hbar/(2\Delta x)$.

Interestingly, as mentioned in the previous chapter, when two operators do commute, they share an eigenbasis, even though their eigenvalues can be completely different.

### 4.1.3 Comments on postulate III

Firstly, this postulate says that it is possible to have multiple outcomes of a measurement for the same system prepared in the same state. This is completely different from classical physics, where two measurements performed on two identical systems will also yield exactly the same result. Secondly, the measurement steers the quantum system toward a particular state, i.e., it alters the state of the system. Again, this is not what happens in classical physics, where, in principle, one can always devise measurements that do not alter in any significant way the state of the system being observed. Thirdly, if we know $|\psi\rangle$ and the spectrum of the operator $\hat{\Omega}$, all we can do as far as predictions of a measurement of $\hat{\Omega}$ are concerned is to compute the probabilities of possible outcomes.

### 4.1.4  Comments on postulate IV

Firstly, notice that Schrödinger's equation is a first-order differential equation. Therefore, given just $|\psi(t = 0)\rangle$ (or at some other reference time), we can in principle determine $|\psi(t)\rangle$ uniquely at any $t > 0$. Secondly, in addition to the initial state, the evolution of the state vector is entirely determined by the Hamiltonian.

But what is a Hamiltonian?

In classical mechanics, a Hamiltonian is a function of the positions and momenta of the particles in the system and is often associated to the system's total energy:

$$H = H(\vec{r}, \vec{p}).$$

It turns out that the evolution of these variables (positions and momenta) are uniquely determined by partial derivatives of the Hamiltonian,

$$\frac{d\vec{p}}{dt} = -\frac{\partial H}{\partial \vec{r}} \quad \text{and} \quad \frac{d\vec{r}}{dt} = \frac{\partial H}{\partial \vec{p}}$$

(these are the so-called Hamilton's equations – an advanced but equivalent formulation of Newtonian mechanics). The Hamiltonian usually consists of a sum of all kinetic and potential energies present in the system, with the latter representing interactions among the system's particles, and between those particles and external fields.

In quantum mechanics, we give the same interpretation to the Hamiltonian, except that we represent it as an operator (or a sum of operators) that encodes the kinetic energies and potential energies in the system. Those energies are written in terms of operators that themselves represent the relevant degrees of freedom of the system.

*Examples of Hamiltonians:*

1. Particle of mass $m$ attached to a spring of constant $k$ and moving in one spatial dimension.

classical Hamiltonian: $H(x,p) = \frac{1}{2m}p^2 + \frac{k}{2}x^2$

quantum Hamiltonian: $\hat{H}(\hat{x},\hat{p}) = \frac{1}{2m}\hat{p}^2 + \frac{k}{2}\hat{x}^2$

Notice that we simply replaced the position and momentum coordinates by their quantum operator counterparts.

2. Magnetic dipole $\vec{\mu}$ in the presence of a uniform magnetic field $\vec{B}$.

   classical Hamiltonian: $H(\vec{\mu}) = -\vec{\mu} \cdot \vec{B}$

   quantum Hamiltonian: $\hat{H}(\widehat{\vec{\mu}}) = -\widehat{\vec{\mu}} \cdot \vec{B}$

   Notice that there is no kinetic term. Moreover, since the magnetic field is fixed (it is not a degree of freedom or variable but rather a parameter) it is not treated as an operator.

Often, it is appropriate to adopt an evolution operator instead of defining the dynamics via Schrödinger's equation. Consider

$$|\psi(t)\rangle = \hat{U}(t;0)|\psi(0)\rangle$$

for $t \geq 0$. The vector on the l.h.s. represents the state at time $t$; the vector on the r.h.s. represents the system at time 0; $\hat{U}(t;0)$ is the operator that takes the system at time 0 and returns it at time $t$. When the Hamiltonian is time independent, one can easily find an expression for $\hat{U}$ from the Schrödinger equation:

$$i\hbar\frac{d}{dt}|\psi(t)\rangle = \hat{H}|\psi(t)\rangle \qquad \longrightarrow \qquad |\psi(t)\rangle = e^{-i\hat{H}t/\hbar}|\psi(0)\rangle$$

(prove it!). We then identify the evolution operator as

$$\hat{U}(t;0) = e^{-i\hat{H}t/\hbar}. \tag{4.3}$$

The evolution operator is an exponential of the Hamiltonian operator!

*Math Digression:* How to define a function of an operator in practice? Use a Taylor series expansion:

$$e^{\hat{A}} = \frac{(\hat{A})^0}{0!} + \frac{(\hat{A})^1}{1!} + \frac{(\hat{A})^2}{2!} + \frac{(\hat{A})^3}{3!} + \ldots = \hat{I} + \hat{A} + \frac{1}{2}\hat{A}^2 + \frac{1}{6}\hat{A}^3 \ldots$$

Integer powers of operators are straightforward to implement: just act with the operator sequentially as many times as the power. Notice that $(\hat{A})^0$ results in the identity operator $\hat{I}$.

### 4.1.5 Comments on postulate V

The concept of a product state that emerges from combining two or more quantum systems into one is hard to grasp but crucial for understanding entanglement. To help a bit, consider as an example a system of two particles of spin $s = 1/2$. In principle, we could prepare the state with both spins up,

$$|\Psi\rangle = |\uparrow\rangle_a \otimes |\uparrow\rangle_b \qquad (4.4)$$

or the state with both spins down,

$$|\Psi\rangle = |\downarrow\rangle_a \otimes |\downarrow\rangle_b.$$

In these two examples, the states of the total system are a product of two separate individual qubit states. We could also prepare a state which is a superposition of these two examples:

$$|\Psi\rangle = \frac{1}{\sqrt{2}}\left(|\downarrow\rangle_a \otimes |\downarrow\rangle_b + |\uparrow\rangle_a \otimes |\uparrow\rangle_b\right). \qquad (4.5)$$

If we consider that we could also prepare up-down and down-up states, which are distinguishable from up-up and down-down, it is clear that the dimension of the Hilbert space for the two-spin system is four.

The state in Eq. (4.5) is particularly interesting because it is an example of an *entangled state*, i.e., a state which cannot be reduced to a single product . Entangled states can only happen in systems composed of two or more subsystems or degrees of freedom (e.g., in this case two qubits). A more intuitive way to think about entanglement has to do with measurement: a state is entangled when measurements of its components yield random but fully correlated results. For instance, for the state in Eq. (4.5), a measurement of qubit $a$ in the $\{\uparrow,\downarrow\}$ basis would yield either $\uparrow$ or $\downarrow$ with equal probability; however, once a certain outcome is obtained, we can be 100% certain that a similar measurement of qubit $b$ will yield exactly the same outcome. The randomness aspect is very important. Take, for example, the state in Eq. (4.4): measurements of qubits $a$ and $b$ also yield exactly the same outcome, but in this case there is no uncertainty in that outcome and therefore the state is not entangled (indeed, it is a single product).

## 4.2 REFERENCES AND FURTHER READING

1. Shankar, R. 1994. *Principles of Quantum Mechanics*, 2nd edition. New York: Plenum Press. Chapter 4.

2. Liboff, R. L. 2003. *Introductory Quantum Mechanics*, 4th edition. Addison Wesley. Chapter 3.

3. Zettili, N. 2009. *Quantum Mechanics, Concepts and Applications*, 2nd edition. John Wiley & Sons. Sections 3.1-3.6.

4. Nielsen M. A. and I. L. Chuang. 2000. *Quantum Computation and Quantum Information*. Cambridge Univ. Press. Section 2.3.

## 4.3  EXERCISES AND PROBLEMS

1. Which of the following states of a system composed of two spin-1/2 particles is entangled? Justify your answer.

   (a) $|\psi_1\rangle = |\uparrow\rangle_a \otimes |\downarrow\rangle_b$

   (b) $|\psi_2\rangle = \frac{1}{\sqrt{2}} \left( |\uparrow\rangle_a \otimes |\downarrow\rangle_b + |\uparrow\rangle_a \otimes |\uparrow\rangle_b \right)$

   (c) $|\psi_3\rangle = \frac{1}{\sqrt{2}} \left( |\uparrow\rangle_a \otimes |\uparrow\rangle_b + |\downarrow\rangle_a \otimes |\downarrow\rangle_b \right)$

   (d) $|\psi_4\rangle = \frac{1}{2} \left( |\uparrow\rangle_a \otimes |\uparrow\rangle_b + |\uparrow\rangle_a \otimes |\downarrow\rangle_b + |\downarrow\rangle_a \otimes |\uparrow\rangle_b + |\downarrow\rangle_a \otimes |\downarrow\rangle_b \right)$

2. Suppose that we take a double-well system and place two electrons on it. Because electrons have spin 1/2, they are "fermions" and two fermions cannot occupy the same quantum state. This result, known as the Pauli principle, forbids, for instance, electrons with the same spin orientation to be in the same orbital configuration. Let us assume that each well can hold only one "energy level", thus forcing two electrons that find themselves in the same well to be on the same energy level but with opposite spin orientations. Because of Coulomb repulsion, we expect a double occupied well to require an extra energy, which we call $U$.

   (a) Find all possible classical configurations that two electrons can take in a double-well system with a single energy level per well (use the notation $|\phi\rangle_L |\psi\rangle_R$ to denote the two-electron state vectors, with suitable choices for $\phi$ and $\psi$, e.g., 0, ↑, ↓, etc.). Associate an orthonormal basis to these configurations.

   (b) Setting the energy of the levels in the wells to $\varepsilon$ and the energy cost of double occupancy to $U$, write down a Hamiltonian operator for this system.

   (c) Now, let tunneling take place, namely, add to the Hamiltonian a term that allows electrons to hop from one well to another (the Dirac notation comes in handy!). Assign a prefactor $t$ to this term in the Hamiltonian.

(d) Find the eigenvalues of the resulting Hamiltonian. *Suggestion*: start with $\varepsilon = 0$ and bring $\varepsilon$ back at the end of your calculation by a suitable shift of the eigenvalues.

(e) What would be an entangled quantum state for this system?

# CHAPTER 5

# Qubits, Gates, and Circuits

Before we start to discuss quantum bits ("qubits" for short), let us spend some time reviewing classical bits, which we will simply call bits, and the types of operations one can perform with them.

## 5.1 CLASSICAL BITS AND LOGIC GATES

Any classical physical system that takes only two clearly distinguishable configurations can become a bit. More abstractly, a bit is any variable that can only take two values.[1] Historically, because bits were implemented with devices running electric currents, we associate to them the values "0" (current off) and "1" (current on). This association is also convenient because any integer or finite-precision real number can be represented using 0s and 1s (the so-called binary decomposition). For instance, for any integer $x$, we can write

$$x = b_{n-1} \times 2^{n-1} + b_{n-2} \times 2^{n-2} + \cdots + b_1 \times 2^1 + b_0 \times 2^0,$$

where $b_k = 0$ or 1 and $k = 0, 1, \ldots, n - 1$. The $b_k$ coefficients are the bit variables. In the expression above, we used $n$ such variables to represent the number $x$. We often line up the bit variables together and in decreasing order from left to right to form a bit string: $b_{n-1}b_{n-2} \cdots b_1 b_0$.

*Examples*:

- 3 (decimal) $= 1 \times 2^1 + 1 \times 2^0 \equiv 11$ (binary)

---

[1] Bit variables are also called Boolean variables.

- 286 (decimal) $= 1 \times 2^8 + 0 \times 2^7 + 0 \times 2^6 + 0 \times 2^5 + 1 \times 2^4 + 1 \times 2^3 + 1 \times 2^2 + 1 \times 2^1 + 0 \times 2^0 \equiv 100011110$ (binary)

We can extend the binary representation to fractional numbers:

$$x = b_{n-1} \times 2^{n-1} + b_{n-2} \times 2^{n-2} + \cdots + b_1 \times 2^1 + b_0 \times 2^0 + f_1 \times 2^{-1} + f_2 \times 2^{-2} + \cdots$$

where $f_k = 0$ or 1. The set $\{f_k\}$ spans the fractional part.

*Example*:

- 3.375 (decimal) $= 1 \times 2^1 + 1 \times 2^0 + 1 \times 2^{-2} + 1 \times 2^{-3} = 11.011$ (binary)

All arithmetic rules we use to manipulate numbers in the decimal representation work for binary numbers, although they look a bit strange:

$0 + 0 = 0$
$0 + 1 = 1$
$1 + 1 = 10$ (we needed to bring a second bit to represent the decimal 2)
$01 + 10 = 11$
$11 + 01 = 100$ (we needed to bring a third bit to represent the decimal 4)
and so on.

## 5.1.1 Circuits

Before we start discussing what we can do with bits, we need to settle on some conventions, including how to graphically represent bits as they undergo transformations and operations. The standard way to do that is to associate to each bit a line, going from left (beginning) to right (end), see Fig. 5.1.

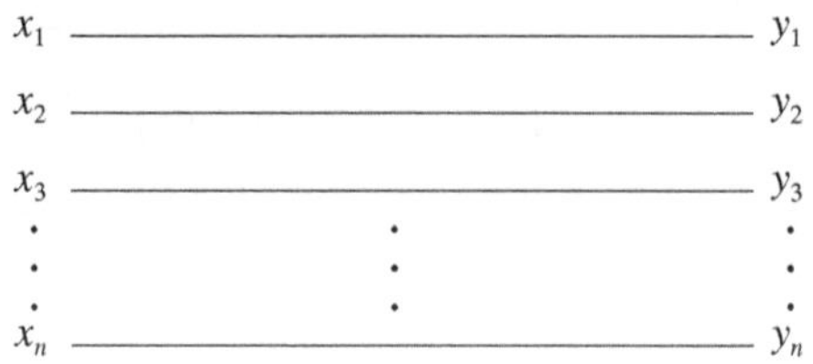

Figure 5.1 Schematic representation of a circuit (empty in this case). Each horizontal line corresponds to the evolution of a binary variable and is called a bitline.

In Fig. 5.1, $\{x_1, x_2, x_3, \ldots, x_n\}$ are the initial values of the bits and $\{y_1, y_2, y_3, \ldots, y_n\}$ are the final ones. We often call "bitlines" the lines representing the time evolution of the bits. Of course, if nothing is acting on the bits, then their initial and final values are the same. When some operation acts on them, we can represent the situation by inserting boxes and other symbols into the diagram to indicate those operations. Say we have an operation that acts on bits 1 and 2, following by another that acts on bit 3, and then another acting on bit $n$; call these operations $A$, $B$, and $C$, respectively. We can represent this sequence of operations as in Fig. 5.2.

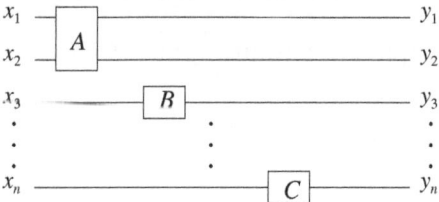

Figure 5.2 Operations on binary variables are indicated by boxes.

We call the sequence of operations a circuit. So, a circuit is basically a chronological sequence of operations. In the classical (non-quantum) case, the operations are called logic gates. (It is a little harder to represent operations that act on non adjacent bitlines, but we will find ways to do it.) Now we are ready to continue!

### 5.1.2 Logic gates

Any complex operation on bits can be broken down into elementary operations that we call logic gates.

To the logic gates we associate truth tables, which are basically lists of all possible input states and the corresponding outputs. For instance, consider the NOT gate:

- NOT: $x$ input; $y$ output: $y = \text{NOT}(x) \equiv \bar{x}$

| $x$ | $y$ |
|---|---|
| 0 | 1 |
| 1 | 0 |

This one-bit gate "flips" the state of the bit.

Other important gates are:

- AND: $x, y$ inputs; $z$ output: $z = \text{AND}(x, y) \equiv x \wedge y = x \cdot y$

| $x$ | $y$ | $z$ |
|---|---|---|
| 0 | 0 | 0 |
| 0 | 1 | 0 |
| 1 | 0 | 0 |
| 1 | 1 | 1 |

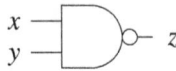

This gate realizes a one-bit multiplication.

- NAND: $x, y$ inputs; $z$ output: $\quad z = \text{NAND}(x, y) \equiv \overline{x \wedge y}$

| $x$ | $y$ | $z$ |
|---|---|---|
| 0 | 0 | 1 |
| 0 | 1 | 1 |
| 1 | 0 | 1 |
| 1 | 1 | 0 |

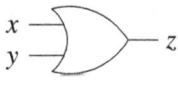

- OR: $x, y$ inputs; $z$ output: $\quad z = \text{OR}(x, y) \equiv x \vee y = x + y$

| $x$ | $y$ | $z$ |
|---|---|---|
| 0 | 0 | 0 |
| 0 | 1 | 1 |
| 1 | 0 | 1 |
| 1 | 1 | 1 |

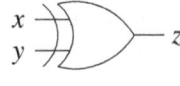

- XOR:[2] $x, y$ inputs; $z$ output: $\quad z = \text{XOR}(x, y) \equiv x \oplus y$

| $x$ | $y$ | $z$ |
|---|---|---|
| 0 | 0 | 0 |
| 0 | 1 | 1 |
| 1 | 0 | 1 |
| 1 | 1 | 0 |

This gate realizes one-bit addition, with a caveat (no carry forward bit).

---

[2]Useful relation: $x \oplus 1 = \bar{x}$.

There are also operations which are not logic gates but perform useful tasks. For instance,

FANOUT: $x$ input; $y$, $z$ outputs: $\quad y = x$, $z = x$

| $x$ | $y$ | $z$ |
|---|---|---|
| 0 | 0 | 0 |
| 1 | 1 | 1 |

It acts like a "copy" gate.

By combining logic gates and operations such as FANOUT, one can build complex operations such as the one shown in Fig. 5.3, which is a one-bit half adder (it is called so because there is no input carry bit). For a full adder, we need to combine two half adders and an OR gate, as shown in Fig. 5.4.

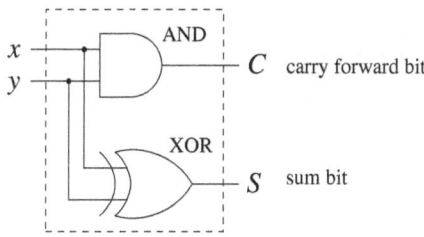

Figure 5.3 A half adder (it is called so because it does not include a carry forward input bit).

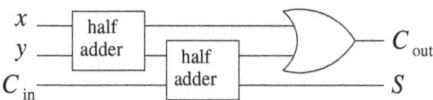

Figure 5.4 A full one-bit adder.

An important aspect of all examples of two-bit gates we presented so far is that they are *irreversible*, namely, you cannot figure out the input state if given only the output state. Check that yourself.

If we consider logic gates as maps (i.e., relations that relate inputs to outputs), we notice that an irreversible two-bit gate $g$ implements the following map:[3]

$$g: \{0,1\}^2 \to \{0,1\}^1.$$

---

[3]Such a map is a Boolean function.

Here, $\{0,1\}^n$ is used as a notation to represent all possible $n$-bit states for a total of $2^n$ configurations. Notice that there is a reduction in the space dimension for gates such as AND, OR, XOR, and NAND, as they take four possible input configurations onto only two (but there is no such a reduction for the one-bit gate NOT). Such a reduction in dimensionality signifies irreversibility.

It turns out that it is possible to accomplish the same logic operation using only reversible gates. However, there is a caveat: sometimes we need to bring fresh bits into the operation, the so-called ancillary bits or ancillae for short, to help. Let us understand how reversible gates work through an example.

*Example*:

Turning an XOR into a CNOT.

- XOR + FANOUT: $x, y$ inputs; $z, w$ outputs

| $x$ | $y$ | $z$ | $w$ |
|---|---|---|---|
| 0 | 0 | 0 | 0 |
| 0 | 1 | 0 | 1 |
| 1 | 0 | 1 | 1 |
| 1 | 1 | 1 | 0 |

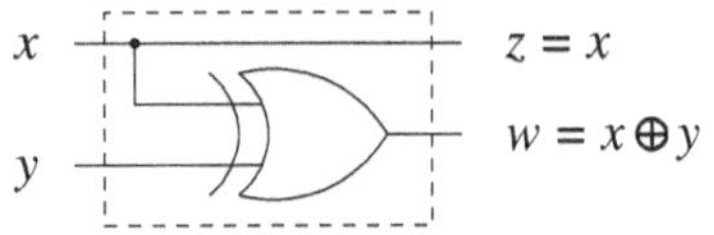

$z = x$

$w = x \oplus y$

Notice that there is a one-to-one correspondence between input states and output states, as illustrated in Fig. 5.5. (In fact, in this particular example, the gate is its own inverse.) By keeping a copy of one of the input bit values, we managed to keep sufficient information about the input state to uniquely determine it from the output. The result is the so-called CNOT gate, which is represented in Fig. 5.6.[4]

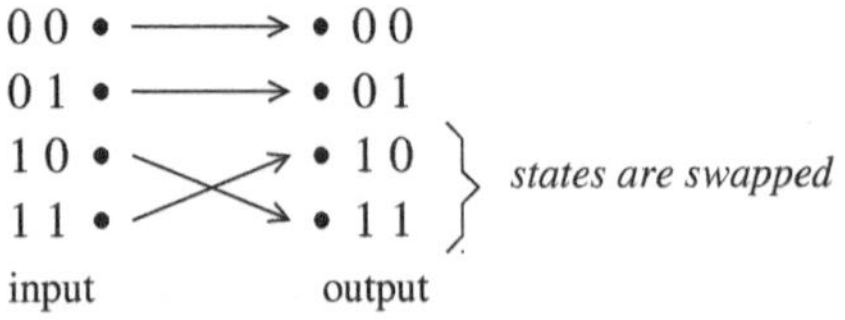

$$00 \bullet \longrightarrow \bullet 00$$
$$01 \bullet \longrightarrow \bullet 01$$
$$\left.\begin{array}{l}10 \bullet \\ 11 \bullet\end{array}\right\} \times \left\{\begin{array}{l}\bullet 10 \\ \bullet 11\end{array}\right\} \text{states are swapped}$$

input        output

Figure 5.5 The CNOT permutation.

---

[4]This is not the best way to create a gate with the CNOT functionality, as it requires the use of two irreversible elements. It is possible and desirable to create CNOT gates without involving irreversible elements.

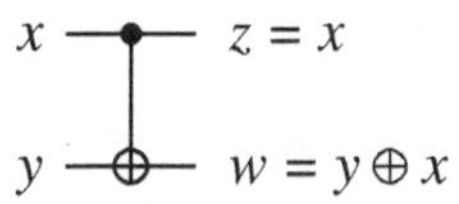

$$x \quad\bullet\quad z = x$$
$$y \quad\oplus\quad w = y \oplus x$$

Figure 5.6 The CNOT gate.

A CNOT is a particular case of a "control gate". The reason for this name is that the value of bit $x$ decides whether the value of bit $y$ is flipped or not on output:

$$x = 0 \quad\longrightarrow\quad w = y = y \oplus 0$$
$$x = 1 \quad\longrightarrow\quad w = \bar{y} = y \oplus 1$$

For this reason, $x$ is called the control bit while $y$ is called the target bit.

Consider now a more general control gate with two control bits, the so-called TOFFOLI gate :

• TOFFOLI: $x_1, x_2, x_3$ inputs; $y_1, y_2, y_3$ outputs

| $x_1$ | $x_2$ | $x_3$ | $y_1$ | $y_2$ | $y_3$ |
|---|---|---|---|---|---|
| 0 | 0 | 0 | 0 | 0 | 0 |
| 0 | 0 | 1 | 0 | 0 | 1 |
| 0 | 1 | 0 | 0 | 1 | 0 |
| 0 | 1 | 1 | 0 | 1 | 1 |
| 1 | 0 | 0 | 1 | 0 | 0 |
| 1 | 0 | 1 | 1 | 0 | 1 |
| 1 | 1 | 0 | 1 | 1 | 1 |
| 1 | 1 | 1 | 1 | 1 | 0 |

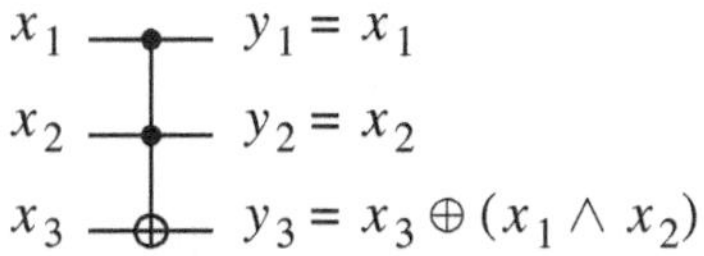

$$x_1 \quad\bullet\quad y_1 = x_1$$
$$x_2 \quad\bullet\quad y_2 = x_2$$
$$x_3 \quad\oplus\quad y_3 = x_3 \oplus (x_1 \wedge x_2)$$

Notice that $x_3$ is flipped only when $x_1 = x_2 = 1$.

## 5.1.3 Universal gate sets

It turns out that the NAND gate is universal, in the sense that any Boolean function $f : \{0,1\}^n \to \{0,1\}^1$ can be decomposed in terms of NAND logic operations.[5] Modern transistor-based processors (like the ones in your computer or cell phone) often rely heavily on NAND gates. Notice that gates such as XOR, AND, and OR are not universal, meaning you cannot decompose a Boolean function using only one type of those gates.

---

[5]FANOUT or copy operations are implicitly assumed to be available.

When it comes to reversible binary operations, the TOFFOLI gate is universal. However, this is only the case if ancillary bits can be utilized and initialized to 0. For some complicated Boolean functions, we may need an exponentially large number of such ancillary bits! More precisely, any Boolean function can be decomposed in terms of TOFFOLI gates, provided that ancillary bits are freely available.

### 5.1.4  Reversible gates and permutations

Because $n$-bit reversible gates implement one-to-one relations between all possible input and output bit strings of length $n$, they represent permutations in the $2^n$-dimensional space of $n$-bit configurations. Such permutations form a group denoted by the symbol $S_{2^n}$.[6] To each reversible gate we can thus associate a permutation, which we represent by a $2^n \times 2^n$ matrix where the elements are either 0 or 1. For instance, consider the CNOT gate $(n = 2)$:

| $x_1$ | $x_2$ | $y_1$ | $y_2$ |
|-------|-------|-------|-------|
| 0 | 0 | 0 | 0 |
| 0 | 1 | 0 | 1 |
| 1 | 0 | 1 | 1 |
| 1 | 1 | 1 | 0 |

$$Y = \begin{pmatrix} 1 & 0 & 0 & 0 \\ 0 & 1 & 0 & 0 \\ 0 & 0 & 0 & 1 \\ 0 & 0 & 1 & 0 \end{pmatrix} \cdot X$$

The input and output states are denoted by 4-entry column vectors $X$ and $Y$, respectively. The correspondence between bit strings and column vectors is the following:

$$00 \to \begin{pmatrix} 1 \\ 0 \\ 0 \\ 0 \end{pmatrix}, \quad 01 \to \begin{pmatrix} 0 \\ 1 \\ 0 \\ 0 \end{pmatrix}, \quad 10 \to \begin{pmatrix} 0 \\ 0 \\ 1 \\ 0 \end{pmatrix}, \quad 11 \to \begin{pmatrix} 0 \\ 0 \\ 0 \\ 1 \end{pmatrix}.$$

Notice that the $4 \times 4$ permutation matrix has an important property: there is only one 1 for every row and column; all other elements are 0.

In the same fashion, we can also associate a $2 \times 2$ permutation matrix to the NOT case:[7, 8]

---

[6]A group is a set whose elements obey certain properties. See https://en.wikipedia.org/wiki/Group_(mathematics) for a definition.

[7]Where have you seen this matrix before? Recall the spin 1/2 quantum systems!

[8]Notice the change in the symbol used to represent a NOT gate. In the context of reversible gates, one adopts this symbol rather than the one used earlier when discussing irreversible gates.

| $x$ | $y$ |
|---|---|
| 0 | 1 |
| 1 | 0 |

$$Y = \begin{pmatrix} 0 & 1 \\ 1 & 0 \end{pmatrix} \cdot X$$

$$x_1 \; \longoplus \; y_1$$
$$\| \qquad\qquad \|$$
$$X \qquad\qquad Y$$

where, in this case, the bit values correspond to the column vectors

$$0 \to \begin{pmatrix} 1 \\ 0 \end{pmatrix} \quad \text{and} \quad 1 \to \begin{pmatrix} 0 \\ 1 \end{pmatrix}.$$

## 5.2  QUBITS AND QUANTUM GATES

Permutations are a special case of a more general transformation called unitary. Unitary transformations can be represented by unitary matrices, which we know are also used to represent unitary operators in quantum mechanics. Thus, there is a close connection between unitary operators and reversible operations. We will elaborate more on that shortly.

If $\hat{P}$ is a permutation matrix,[9] it is always possible to obtain its inverse $\hat{P}^{-1}$,

$$\hat{P} \cdot \hat{P}^{-1} = \hat{I},$$

where $\hat{I}$ is the identity matrix, containing zeros everywhere except along the diagonal, which has only 1s. Guess what is $\hat{P}^{-1}$? The transpose of $\hat{P}$, namely, $\hat{P}^T$ (transposition means replacing columns by rows, and vice versa).

We have already seen matrices with a similar property: the unitary ones. If $\hat{U}$ is a unitary matrix, then $\hat{U}^{-1} = \hat{U}^\dagger$, where $\hat{U}^\dagger = (\hat{U}^T)^*$.

The main difference between permutation and unitary matrices is that permutation matrix elements are either 0 or 1, and there is always only one 1 per row or column. Unitary matrix elements are not restricted to only 0s and 1s, and not even to real numbers.

While permutation transformations represent reversible operations among bits, unitary operations are used to implement quantum operations among qubits. Let us begin to understand this concept through examples.

*Examples*:

1. Consider the unitary matrix

$$\hat{U} = \begin{pmatrix} 1 & 0 \\ 0 & -1 \end{pmatrix},$$

---

[9]To distinguish matrices from scalars, we use a hat, similarly (and not accidentally!) to what we did for quantum operators.

which is not a permutation (notice the element with a minus sign). Then, let $|x\rangle$ and $|y\rangle$ represent input and output qubit states, such that

$$|y\rangle = \hat{U}|x\rangle.$$

Notice that if

$$|x\rangle = \begin{pmatrix} 0 \\ 1 \end{pmatrix},$$

then

$$|y\rangle = \begin{pmatrix} 0 \\ -1 \end{pmatrix} = (-1) \times \begin{pmatrix} 0 \\ 1 \end{pmatrix}.$$

The unitary transformation returned the input vector multiplied by the scalar $-1$. The latter is a phase factor: $-1 = e^{i\pi}$. No classical transformation (i.e., a sequence of logic gates) can add a phase factor to a qubit state vector. Thus $\hat{U}$ has no counterpart in classical computing.

Let $|0\rangle$ denote the state where the qubit is "0" and $|1\rangle$ when it is "1". Then,

$$\hat{U}|0\rangle = |0\rangle \quad \text{and} \quad \hat{U}|1\rangle = -|1\rangle.$$

2. Consider a more general phase gate, defined by the matrix

$$\hat{U} = \begin{pmatrix} 1 & 0 \\ 0 & e^{i\theta} \end{pmatrix}.$$

In this case,

$$\hat{U}|0\rangle = |0\rangle \quad \text{and} \quad \hat{U}|1\rangle = e^{i\theta}|1\rangle.$$

All classical *reversible* gates are also quantum gates, in the sense that any permutation is a unitary transformation as well.

Is it possible to turn any irreversible gate into a quantum one? The generic answer is no. The physical reason is that the number of "bit lines", or degrees of freedom, is not preserved in an irreversible gate. Irreversible gates do not preserve the dimension of the qubit space. In a quantum evolution, degrees of freedom are preserved, namely, they cannot be destroyed or created. There is a surprising consequence to that, which we will study soon (*spoiler*: copy via FANOUT is not allowed in quantum information processing).

Hence, the only similarity between qubits and classical bits is that the former can only be measured or observed in two states. Qubits go much beyond classical bits. Qubits can exist in many different states, infinitely many, in fact. Consider the qubit state

$$|\psi\rangle = a|0\rangle + b|1\rangle,$$

which represents a superposition between "0" and "1". Because $a$ and $b$ can vary continuously (and can even be complex numbers), there is an infinite number of possibilities for $|\psi\rangle$.

### 5.2.1 Single-qubit gates

There are a few single-qubit gates that are particularly useful and important. Below, we represent the unitary operator associated to each gate as a matrix in the basis provided by the "0" and "1" states, namely,

$$|0\rangle = \begin{pmatrix} 1 \\ 0 \end{pmatrix} \quad \text{and} \quad |1\rangle = \begin{pmatrix} 0 \\ 1 \end{pmatrix}.$$

- NOT:

$$\hat{U}_{\text{NOT}} = \begin{pmatrix} 0 & 1 \\ 1 & 0 \end{pmatrix}$$

- Phase:

$$\hat{U}_\theta = \begin{pmatrix} 1 & 0 \\ 0 & e^{i\theta} \end{pmatrix}$$

There are two special phase gates worth mentioning :

$$\hat{S} = \hat{U}_{\pi/2} = \begin{pmatrix} 1 & 0 \\ 0 & i \end{pmatrix} \quad \text{and} \quad \hat{T} = \hat{U}_{\pi/4} = \begin{pmatrix} 1 & 0 \\ 0 & e^{i\pi/4} \end{pmatrix}.$$

- Hadamard:

$$\hat{U}_H = \frac{1}{\sqrt{2}} \begin{pmatrix} 1 & 1 \\ 1 & -1 \end{pmatrix}$$

This one is especially useful for creating superposition states . For instance,

$$\hat{U}_H|0\rangle = \frac{1}{\sqrt{2}}(|0\rangle + |1\rangle)$$

$$\hat{U}_H|1\rangle = \frac{1}{\sqrt{2}}(|0\rangle - |1\rangle)$$

- Pauli: there are three of them:

$$\hat{X} = \begin{pmatrix} 0 & 1 \\ 1 & 0 \end{pmatrix}$$

$$\hat{Y} = \begin{pmatrix} 0 & -i \\ i & 0 \end{pmatrix}$$

$$\hat{Z} = \begin{pmatrix} 1 & 0 \\ 0 & -1 \end{pmatrix}$$

Notice that $\hat{X}$ is the same as the NOT gate and $\hat{Z}$ is the same as the phase gate with $\theta = \pi$. Pauli gates have some intriguing properties:

$\hat{X}^2 = \hat{Y}^2 = \hat{Z}^2 = \hat{I}$ (even powers of Pauli operators are the identity)

$\hat{X}\hat{Y} = i\hat{Z}, \quad \hat{Z}\hat{X} = i\hat{Y}, \quad \hat{Y}\hat{Z} = i\hat{X}$ (cyclic product permutation rule)

$[\hat{X}, \hat{Y}] = 2i\hat{Z}, \quad [\hat{Z}, \hat{X}] = 2i\hat{Y}, \quad [\hat{Y}, \hat{Z}] = 2i\hat{X}$ (they form a so-called Lie algebra).

Aside their use as quantum gates, Pauli operators are important in other areas of quantum mechanics and used to represent observables such as the components of the $s = 1/2$ spin. Spin is an intrinsic angular momentum of elementary particles, such as the electron.[10]

---

[10]The $x$, $y$, and $z$ components of the angular momentum do not commute and therefore cannot be simultaneously determined with certainty through measurements. The Pauli operators manifest this property via their non commutability.

- Rotations:

$$-\boxed{R_x(\theta)}-$$

$$\hat{R}_x(\theta) = e^{-i\theta\hat{X}/2} = \cos(\theta/2)\hat{I} - i\sin(\theta/2)\hat{X}$$

$$-\boxed{R_y(\theta)}-$$

$$\hat{R}_y(\theta) = e^{-i\theta\hat{Y}/2} = \cos(\theta/2)\hat{I} - i\sin(\theta/2)\hat{Y}$$

$$-\boxed{R_z(\theta)}-$$

$$\hat{R}_z(\theta) = e^{-i\theta\hat{Z}/2} = \cos(\theta/2)\hat{I} - i\sin(\theta/2)\hat{Z}$$

In matrix representation,

$$\hat{R}_x(\theta) = \begin{pmatrix} \cos(\theta/2) & -i\sin(\theta/2) \\ -i\sin(\theta/2) & \cos(\theta/2) \end{pmatrix}$$

$$\hat{R}_y(\theta) = \begin{pmatrix} \cos(\theta/2) & -\sin(\theta/2) \\ \sin(\theta/2) & \cos(\theta/2) \end{pmatrix}$$

$$\hat{R}_z(\theta) = \begin{pmatrix} e^{-i\theta/2} & 0 \\ 0 & e^{i\theta/2} \end{pmatrix}.$$

It is possible to prove that any one-qubit unitary operator (i.e., any one-qubit gate) can be decomposed as

$$e^{i\alpha}\hat{R}_z(\beta)\hat{R}_y(\gamma)\hat{R}_z(\delta),$$

where $\alpha$, $\beta$, $\gamma$, and $\delta$ are real numbers. We will return to this topic (gate decomposition) later on. For now, notice that four real parameters $(\alpha, \beta, \gamma, \delta)$ are needed in the most general case, which matches the number of independent real numbers needed to write the most general $2 \times 2$ unitary matrix.

## 5.3 BLOCH SPHERE REPRESENTATION

Let us introduce a neat way to represent single-qubit states: the Bloch sphere.

The most general state vector of a single qubit can be written in the so-called computational basis (i.e., the basis of "0" and "1") as

$$|\psi\rangle = \begin{pmatrix} c_1 \\ c_2 \end{pmatrix} = c_1|0\rangle + c_2|1\rangle,$$

where $c_1$ and $c_2$ are complex numbers satisfying $|c_1|^2 + |c_2|^2 = 1$. Let $c_1 = \alpha_1 + i\beta_1$ and $c_2 = \alpha_2 + i\beta_2$, where $\alpha_1$, $\alpha_2$, $\beta_1$, and $\beta_2$ are real numbers. In terms of these numbers, the normalization condition reads

$$\alpha_1^2 + \alpha_2^2 + \beta_1^2 + \beta_2^2 = 1 \tag{5.1}$$

(therefore, $|\alpha_1| \leq 1$, $|\alpha_2| \leq 1$, $|\beta_1| \leq 1$, and $|\beta_2| \leq 1$). Mathematically, this equation describes the surface of a sphere of radius 1 in a four-dimensional space. Thus we only need three coordinates to uniquely determine any point on such a surface embedded in four dimensions. As a result, in principle we can map the coefficients $\alpha_1$, $\alpha_2$, $\beta_1$, and $\beta_2$ under the constraint of Eq. (5.1) onto three coordinates $(x, y, z)$. The mapping is not unique, but consider for instance this one:

$$
\begin{aligned}
x &= 2(\alpha_1\alpha_2 + \beta_1\beta_2) \\
y &= 2(\beta_2\alpha_1 - \alpha_2\beta_1) \\
z &= \alpha_1^2 + \beta_1^2 - \alpha_2^2 - \beta_2^2.
\end{aligned}
$$

The convenience of this particular map is that it is consistent with the relation

$$x^2 + y^2 + z^2 = 1$$

(check it!). Therefore, the coordinates $(x, y, z)$ themselves describe points on a *three-dimensional* sphere of radius 1, the so-called Bloch sphere. We can express the points on the Bloch sphere via spherical coordinates $(r, \theta, \phi)$ by setting $r = 1$ and

$$
\begin{aligned}
x &= \sin\theta\cos\phi \\
y &= \sin\theta\sin\phi \\
z &= \cos\theta,
\end{aligned}
$$

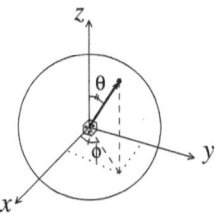

with $0 \leq \theta \leq \pi$ and $0 \leq \phi \leq 2\pi$. Notice that through these mappings we went from four variables to only two angles. Going back to the original four variables, we find

$$
\begin{aligned}
\alpha_1 &= \cos(\theta/2)\cos(\omega) \\
\alpha_2 &= \sin(\theta/2)\cos(\omega + \phi) \\
\beta_1 &= \cos(\theta/2)\sin(\omega) \\
\beta_2 &= \sin(\theta/2)\sin(\omega + \phi),
\end{aligned}
$$

where $\omega$ is some unspecified angle, with $0 \leq \omega \leq 2\pi$. This angle or phase has no physical significance. To understand why, let us go back to the matrix and braket representations. We find that (check this too!)

$$|\psi\rangle = e^{i\omega} \begin{pmatrix} \cos(\theta/2) \\ e^{i\phi}\sin(\theta/2) \end{pmatrix} = e^{i\omega}[\cos(\theta/2)|0\rangle + e^{i\phi}\sin(\theta/2)|1\rangle]. \quad (5.2)$$

Notice that the factor $e^{i\omega}$ is an overall phase and, as such, cannot be detected. We henceforth disregard it. As a result, by convention, in the Bloch representation, the cofficient in front of the $|0\rangle$ basis state must always be real and positive.

To recap: one can go from a four-dimensional to a three-dimensional representation of qubit states, where to each state we associate a point on a sphere of radius 1. While the $(x, y, z)$ coordinates of points on the sphere do not have anything to do with real-space coordinates, it turns out that they can help us visualize qubit states as if they were standard three-dimensional vectors. Thus the great appeal of the Bloch representation.

## 5.3.1 States and gates in the Bloch representation

Notice that at $\theta = 0$ (i.e., at the north pole in the Bloch sphere), we obtain

$$|\psi\rangle = \begin{pmatrix} 1 \\ 0 \end{pmatrix} = |0\rangle,$$

whereas at $\theta = \pi$ (at the south pole), we obtain instead

$$|\psi\rangle = \begin{pmatrix} 0 \\ e^{i\phi} \end{pmatrix} = e^{i\phi}|1\rangle.$$

For $\theta = \pi/2$ (at the equator), we obtain

$$|\psi\rangle = \begin{pmatrix} 1/\sqrt{2} \\ e^{i\phi}/\sqrt{2} \end{pmatrix} = \frac{1}{\sqrt{2}}(|0\rangle + e^{i\phi}|1\rangle),$$

which is a superposition state. Notice that by varying $\phi$, one can move along the equator. Hence, $\theta$ is a latitude angle, while $\phi$ is an azimuth or longitude angle.

What happens when you act with a rotation operator on a state represented on a Bloch sphere? Let us look at a rotation by an angle $\alpha$ around the "$x$ axis", starting from the north pole:

$$\hat{R}_x(\alpha)|0\rangle = [\cos(\alpha/2)\hat{I} - i\sin(\alpha/2)\hat{X}]|0\rangle = \cos(\alpha/2)|0\rangle - i\sin(\alpha/2)|1\rangle.$$

Setting $\alpha = \pi$, we get

$$\hat{R}_x(\pi)|0\rangle = -i|1\rangle,$$

which essentially flips the qubit state by taking it from the north to the south pole, up to a phase factor. When we set $\alpha = \pi/2$ instead, we get

$$\hat{R}_x(\pi/2)|0\rangle = \frac{1}{\sqrt{2}}(|0\rangle - i|1\rangle),$$

which is a state on the equator line. As you can see, a 90° rotation around the $x$ axis bring the qubit state from the north pole to the equator. This aligns perfectly with the geometry afforded by the Bloch sphere representation. The Bloch sphere notation provides an intuitive, visually appealing way of representing qubit states. But there are some subtleties. Let us consider a rotation by $\theta = 2\pi$ of a generic state:

$$
\begin{aligned}
\hat{R}_x(2\pi)|\psi\rangle &= \begin{pmatrix} e^{-i\pi} & 0 \\ 0 & e^{i\pi} \end{pmatrix} \begin{pmatrix} \cos(\theta/2) \\ e^{i\phi}\sin(\theta/2) \end{pmatrix} \\
&= \begin{pmatrix} -\cos(\theta/2) \\ -e^{i\phi}\sin(\theta/2) \end{pmatrix} = -|\psi\rangle.
\end{aligned}
$$

We do not quite get back to the same state we started with! The sign changes! In fact, we need to rotate the state by $\theta = 4\pi$ to get back to exactly the same state we started with:

$$\hat{R}_x(4\pi)|\psi\rangle = |\psi\rangle.$$

Dirac found an entertaining demonstration of this effect: the Dirac string trick! If you search it on the Internet, you will find videos showing it (you can also search it as the plate or belt trick). Technically, it means that the mathematical group of quaternions (our $2 \times 2$ single-qubit unitary operations), known as SU(2), covers the group of three-dimensional rotations, known as SO(3), twice over.

Let us look at the effects of a couple more gates on states on the Bloch sphere:

- Hadamard gate:

$$\hat{H}|0\rangle = \frac{1}{\sqrt{2}}|0\rangle + \frac{1}{\sqrt{2}}|1\rangle,$$

which implies $\theta = \pi/2$ and $\phi = 0$ (the point at the intersection between the $x$ axis and the equator).

- $S$ gate:

$$\hat{S}\left(\frac{1}{\sqrt{2}}|0\rangle + \frac{1}{\sqrt{2}}|1\rangle\right) = \frac{1}{\sqrt{2}}|0\rangle + \frac{i}{\sqrt{2}}|1\rangle,$$

which implies $\theta = \pi/2$ (no change on this angle) and $\phi = \pi/2$ (the point at the intersection between the $y$ axis and the equator).

## 5.4 PAULI DECOMPOSITION

Any one-qubit operator can be decomposed as a weighed sum of Pauli operators complemented by the identity operator:

$$\hat{O} = \begin{pmatrix} O_{11} & O_{12} \\ O_{21} & O_{22} \end{pmatrix} = c_I\,\hat{I} + c_X\,\hat{X} + c_Y\,\hat{Y} + c_Z\,\hat{Z}, \qquad (5.3)$$

where $c_I, c_X, c_Y$, and $c_Z$ are suitable complex coefficients (try to find how they related to the original matrix elements $O_{ij}$). This should not come as a surprise: any one-qubit operator can be written as a $2 \times 2$ matrix with four matrix elements. So, there are four complex degrees of freedom in expressing a one-qubit operator. Equation (5.3) is basically a linear mapping from four complex coefficients onto four others. It can also be understood as a basis decomposition: the Pauli matrices, complemented with the identity, are linearly independent (i.e., we cannot write one in terms of the three others), thus they form a basis in the vector space of $2 \times 2$ complex matrices.

## 5.5 GATES AS PULSES

How does one apply a quantum gate in practice? The action of a gate is performed by applying fields to the physical system that encodes the qubits. Consider for instance the $\hat{R}_x(\alpha)$ gate. It can be implemented by turning on a field[11] whose effect is to add the following term to the qubit Hamiltonian:

$$\hat{H} = \lambda\,\hat{X},$$

where $\lambda$ is the coupling amplitude. If this is the only term in the qubit Hamiltonian, recalling Eq. (4.3), we can easily obtain the evolution operator for the qubit under the action of the field:

$$\hat{U}(t) = e^{-i\hat{H}t/\hbar} = e^{-i(\alpha/2)\hat{X}},$$

---

[11]The field could be electric, magnetic, optical, or a combination of.

where $\alpha = 2\lambda t/\hbar$. Hence, but adjusting the duration of the field pulse, we can control the parameter $\alpha$. For instance, setting $t = \pi\hbar/2\lambda$ creates an $\hat{R}_x(\pi)$ gate. In this context, such a gate is called a $\pi$ pulse.

Similarly, one can implement an $\hat{R}_x(\pi/2)$ gate – which is called a $\pi/2$ pulse – or any other rotation by applying a suitable field pulse and controlling its duration. Other single-qubit gates can be performed in a similar fashion but may require a combination of fields to achieve the correct phases and amplitudes in the qubit state vector. Even two-qubit gates (to be explained in following sections) also boil down to applying a sequence of field pulses to the qubits involved.

## 5.6  MEASUREMENTS

We have not talked much about measurements so far but they are a critical part of any quantum information processing. After all, at some point in the processing, being it computation, storage, communication, or sensing, we need to retrieve information.

Recall that Postulate III of quantum mechanics states that any time we measure an observable in a quantum system, we only obtain one of the eigenvalues of the operator associated to that observable. Moreover, the state of the system after the measurement is that of the eigenvector associated to the observed eigenvalue.

These rules are satisfied by qubits, of course. When we measure some observable involving a certain number of qubits, those qubits will collapse to the eigenstate corresponding to the measured outcome.

Often, we do not specify very clearly which quantity or observable we are measuring in a qubit system; when this is the case, it is implicitly assumed that we are measuring the $\hat{Z}$ operator associated to the qubits. For instance, in a 10-qubit system, measuring qubit #3 means measuring the value of the observable $Z$ associated to that qubit. The eigenvalues of the $\hat{Z}$ operators are $\pm 1$, but we usually prefer to associate them to 0 and 1, (0 for +1 and 1 for −1). To be more precise, when we are expecting 0 or 1 as the outcomes of a qubit measurement, in reality we are measuring the operator $\hat{Q} = (\hat{I} - \hat{Z})/2$. It is trivial to check that $\hat{Q}$ has the following form in the computational basis:

$$\hat{Q} = \begin{pmatrix} 0 & 0 \\ 0 & 1 \end{pmatrix}.$$

In circuits, we denote the "measurement of a qubit" (i.e., measuring the observable $Q$) by the element shown in Fig. 5.7.

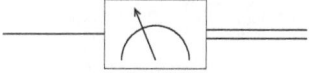

Figure 5.7  Graphic representation of a measurement circuit element.

After measuring a qubit and, say, obtaining 0, that qubit collapses to the state $|0\rangle$, which is in fact a classical state. Similarly, if we obtain 1, the qubit collapses to the state $|1\rangle$. Because the output of a measurement is a classical variable (with values 0 or 1 in this case) and the final state of the qubit is, in essence, classical (i.e., not a superposition), the output line in the measurement element is doubled to contrast with single "quantum" lines.

## 5.7  PRACTICING WITH QISKIT

Let us practice what we learned so far by using Qiskit to play with qubits and quantum gates.

Follow the instruction provided in Appendix B to install Qiskit on your computer and get acquainted with its basic commands. Then try the tasks listed below, making sure to understand the result of each gate application.

1. Create a single-qubit circuit with the qubit set initially to $|0\rangle$; apply a Hadamard gate; print out the resulting state vector.

2. Add a measurement to the circuit and use a sampling evaluator (mimicking a real quantum computer) to run the circuit 1000 times and collect the measurement data.

3. Plot a histogram with the number of counts for each measurement outcome.

4. Plot a vector of Cartesian coordinates $(0, 1, 0)$ on the Bloch sphere.

5. Plot a vector of spherical coordinates $(1, \pi/2, \pi/3)$ on the Bloch sphere.

6. Define another single-qubit circuit with the qubit initially set to $|0\rangle$; apply the sequence of gates Hadamard, $X$, and $Y$, and plot the state vector on the Bloch sphere after each gate.

7. Apply an $R_x(-\pi/2)$ gate. Where is the resulting state vector located on the Bloch sphere?

8. Where does the state vector go after applying an $R_y(\pi/2)$ gate?

## 5.8 TWO- AND MULTI-QUBIT GATES

We have already encountered a two-qubit gate: the CNOT. It applies a NOT gate on one of the qubits (the target one) when the state of the other qubit (the control one) is 1. While not necessarily a quantum gate (it can also be used in classical reversible computations), the CNOT has an entangling effect when applied to qubits in a superposition state. Let us understand this important point.

Consider the initial two-qubit state where the control qubit is in a superposition and the target qubit is in the $|0\rangle$ state:

$$|\psi\rangle = \frac{1}{\sqrt{2}}(|0\rangle + |1\rangle) \otimes |0\rangle = \frac{1}{\sqrt{2}}(|00\rangle + |10\rangle).$$

Notice that this state is fully factorizable and therefore not entangled. Let us apply a CNOT operator to it:

$$\hat{U}_{\text{CNOT}}|\psi\rangle = \frac{1}{\sqrt{2}}(|00\rangle + |11\rangle).$$

Clearly, the resulting state is entangled (i.e., it is not factorizable). Adopting a two-qubit computational basis where

$$|00\rangle = \begin{pmatrix} 1 \\ 0 \\ 0 \\ 0 \end{pmatrix}, \quad |01\rangle = \begin{pmatrix} 0 \\ 1 \\ 0 \\ 0 \end{pmatrix}, \quad |10\rangle = \begin{pmatrix} 0 \\ 0 \\ 1 \\ 0 \end{pmatrix}, \quad \text{and } |11\rangle = \begin{pmatrix} 0 \\ 0 \\ 0 \\ 1 \end{pmatrix},$$

and taking the qubit on the left within the ket as the control, we can represent the CNOT gates in a matrix form,

$$\hat{U}_{\text{CNOT}} = \begin{pmatrix} 1 & 0 & 0 & 0 \\ 0 & 1 & 0 & 0 \\ 0 & 0 & 0 & 1 \\ 0 & 0 & 1 & 0 \end{pmatrix}.$$

The CNOT gate is a particular example of a more general class of $n$-qubit gates named controlled-$U$ gates, as shown in Fig. 5.8.

The gate action is defined by the rule:

- if the control qubit $= |0\rangle$, do nothing;

- if the control qubit $= |1\rangle$, apply the unitary operator $\hat{U}$ on the target bits.

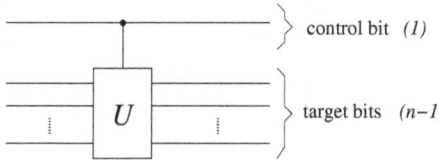

**Figure 5.8** A generic control-$U$ gate.

Here, $\hat{U}$ is a multi-qubit unitary operator acting on a $2^{n-1}$-dimensional space. For instance, for a CNOT, $n = 2$ and $\hat{U} = \hat{X}$.[12]

It turns out that one can build multi-qubit control gates out of CNOT gates. Thus, it is often sufficient to consider circuits with only two-qubit control gates. Let us elaborate on this point.

Consider that any one-qubit unitary operator can be decomposed as a sequence of rotations and an overall phase factor,

$$\hat{U} = e^{i\alpha}\hat{R}_z(\beta)\hat{R}_y(\gamma)\hat{R}_z(\delta) \tag{5.4}$$

after suitable choices for the angles $\alpha$, $\beta$, $\gamma$, and $\delta$. One can show after a bit of algebra that this result can be recast as

$$\hat{U} = e^{i\alpha}\hat{A}\hat{X}\hat{B}\hat{X}\hat{C},$$

where we introduced the following combinations of operators:

$$
\begin{aligned}
\hat{A} &= \hat{R}_z(\beta)\hat{R}_y(\gamma/2) \\
\hat{B} &= \hat{R}_y(-\gamma/2)\hat{R}_z(-(\delta+\beta)/2) \\
\hat{C} &= \hat{R}_z((\delta-\beta)/2).
\end{aligned}
$$

Interestingly, $\hat{A}\hat{B}\hat{C} = \hat{I}$ and $\hat{C}\hat{B}\hat{A} = \hat{I}$. We can imagine that $\hat{U}$ acts on a target qubit while another qubit (the control) activates the $\hat{X}$ operators, as shown in the circuit in Fig. 5.9.

Notice the insertion of a phase gate on the control bitline to account for the overall phase factor in Eq. (5.4).

We can also define control circuits, which we denote by $C_U$. Consider a multi-qubit unitary operator $\hat{U}$ that consists of a series of individual gates, as shown in Fig. 5.10.

A possible $C_U$ circuit is shown in Fig. 5.11.

---

[12]We can write an explicit expression for the control-$U$ gate as $\hat{C}_U = |0\rangle\langle0| \otimes \hat{I}_{n-1} + |1\rangle\langle1| \otimes \hat{U}$, where the projectors act on the control bit and both $\hat{I}_{n-1}$ and $\hat{U}$ act on $n - 1$ target bits.

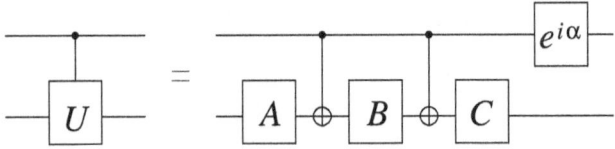

Figure 5.9 The decomposition of a two-qubit control-$U$ gate.

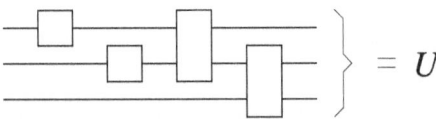

Figure 5.10 A multi-qubit circuit can be cast as a unitary operator $U$.

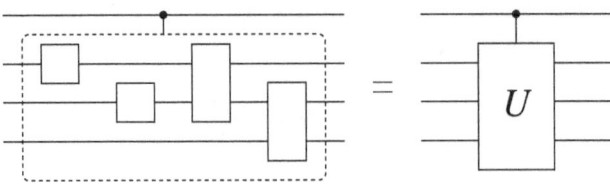

Figure 5.11 A $U$-circuit control gate.

Because each gate in the sub-circuit represented by $\hat{U}$ only acts when the control bit activates them, we can assign individual control lines to each gate making up $\hat{U}$, as shown in Fig. 5.12.

This example illustrates that more complex control-like circuits can be easily decomposed into elementary control gates. Therefore, we can build very complex quantum circuits using just single-qubit gates and CNOT gates. But is that sufficient? We address this question in Sec. 5.10.

## 5.9 MORE PRACTICE WITH QISKIT

Let us practice with one- and two-qubit gates using Qiskit . Try the tasks listed below, always making sure to understand the result of each gate application.

1. Create a two-qubit circuit with all qubits initially set to $|0\rangle$; apply a Hadamard gate on the first qubit and then a CNOT gate with the control on the first qubit and the target on the second qubit; print out the resulting state vector. What kind of state is it?

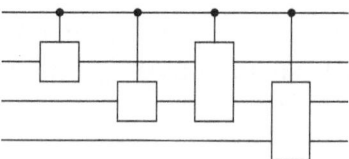

Figure 5.12  A circuit of control gates equivalent to that in Fig. 5.11.

2. Retrieve the unitary operator associated to the circuit and print out the corresponding $4 \times 4$ matrix (in the computational basis).

3. Now add measurements to both qubits; use a sampling evaluator to run the circuit 1000 times and collect the measurement data; plot a histogram with the number of counts for each measurement outcome. Does the result match what you expected?

4. Create a new two-qubit circuit, again with all qubits set initially to $|0\rangle$ again; apply the same gates as in the previous circuit (do not include the measurements yet); in addition, apply a Hadamard gate to the second qubit and then another CNOT gate, this type swapping the location of control and target; print out the resulting state vector. What kind of state is it?

5. Repeat steps 2 and 3 for the new circuit. Reflect on the result of step 3.

## 5.10   UNIVERSAL QUANTUM GATE SETS

What is the smallest set of gates needed to allow for any quantum computation, no matter how complex? In other words: what is the minimum gate set that is needed for implementing any unitary operation over $n$ qubits?

The answer depends on how accurate we want the computation to be. Since unitary operations are defined in the continuum, unless the operation is relatively simple, we can anticipate that a large number of individual elementary gates will be needed to perform complex, multi-qubit operations. Yet, perhaps all that is needed is a good-enough approximation to the unitary operation rather than an exact implementation. In this case, the number of elementary gates may scale more favorably with the number of qubits (i.e., the dimension of the Hilbert space), or, alternatively, only a few types of elementary gates may be sufficient. Let us formalize this notion.

Let $\hat{U}$ be a desired unitary operation and $\hat{V}$ some other unitary operation. We say that $\hat{V}$ approximates $\hat{U}$ to accuracy $\epsilon$ when

$$\max_{|\psi\rangle} \|(\hat{U} - \hat{V})|\psi\rangle\| < \epsilon$$

for $\epsilon > 0$.[13] In words, the maximum value of the norm of the difference between the vectors resulting from acting with $\hat{U}$ and $\hat{V}$ on $|\psi\rangle$ is smaller than a certain desired accuracy. The smaller the $\epsilon$ the better $\hat{V}$ approximates $\hat{U}$. The hope is that $\hat{V}$ can be built out of a small set of gates types (although we may need to use the same type of gate many times). When $\epsilon$ can be made arbitrarily small, we say that $\hat{V}$ approximates $\hat{U}$ with *arbitrary* accuracy.

A set of gates is called universal if, for any $n \geq 1$, any $n$-qubit unitary operator can be approximated with arbitrary accuracy by a quantum circuit formed only by gates from that set.

Based on these definitions, it is possible to prove the following theorem:

*A set composed of any two-qubit entangling gate, together with all one-qubit gates, is universal for quantum computing.*

But what is an entangling gate? It is any gate that, given a product state as input, can output an entangled state (i.e., a state that cannot be factorized, see the beginning of Sec. 5.8). CNOT is an entangling gate; however, the SWAP gate is not (see Sec. 5.11 below). Thus, by combining two-qubit CNOT gates with one-qubit gates one can build any multi-qubit unitary operation.

Is it possible to restrict the one-qubit gates to a smaller set?

The answer is yes. For instance, if you recall, using a series of rotation gates such as $\hat{R}_z$ and $\hat{R}_y$ we can emulate any one-qubit unitary operator, as long as we can freely adjust the rotation angles.

Can we restrict the one-qubit gate class even further? The answer is again positive: the set $\{\hat{H}, \hat{T}\}$ combined with CNOT, where $\hat{H}$ is the Hadamard gate and $\hat{T}$ is the $\pi/4$ phase gate, is universal. This means that any one-qubit unitary operator can be approximated with arbitrary accuracy by a sufficiently large number of $\hat{H}$ and $\hat{T}$ gates.

---

[13]This notion means: the maximum of magnitude of the state $(\hat{U} - \hat{V})|\psi\rangle$ with respect to all possible $|\psi\rangle$ must be smaller than $\epsilon$.

$\{\hat{H}, \hat{T}, \text{CNOT}\}$ is not the only universal set. There are others, but they all are similar in the sense that they contain at least one entangling two-qubit gate and a finite number of one-qubit gate types that can be composed to represent any one-qubit unitary operations with arbitrary accuracy.[14]

The fact that a universal set can approximate any multi-qubit unitary operation with arbitrary accuracy tells us nothing about how efficient such an approximation might be, namely, how many individual gates are needed. When an exponential number of gates is needed, the concept of a universal set is not practical.

Fortunately, a theorem by Solovay and Kitaev saves the day:

*Any one-qubit unitary operator can be approximated with an error $\epsilon$ using at most $O(\log^c(1/\epsilon))$ gates from a set $G$ if, for any gate $g \in G$ its inverse $g^{-1}$ can be implemented exactly by a finite number of gates from $G$.*

Notice that poly-log scaling grows much slower than an exponential one.

Luckily, the subset $\{\hat{H}, \hat{T}\}$ satisfies this condition! However, if we were to substitute $\hat{T}$ by, say, the phase gate $\hat{S}$ the subset would no longer be universal for all one-qubit unitary operations. Nevertheless, the set $\{\text{CNOT}, \hat{H}, \hat{S}\}$ is of some importance in QIP since it generates the so-called Clifford group. A theorem by Gottesman and Knill states that any quantum circuit based on gates from the Clifford group can be efficiently simulated on a classical computer. Namely, it is possible to mimic the action of a Clifford gate using only a polynomial amount of classical computational resources. Because Clifford gates can entangle, this theorem basically says that not all entangled states are difficult to simulate in a classical computer. Clifford gates have applications in certain protocols of entanglement distillation and quantum error correction, as we will see later on. But they are insufficient for universal quantum computation.

## 5.11  OTHER GATES OF RELEVANCE

To wrap up this chapter, let us describe a few more two-qubit gates of relevance, with their symbols and matrix representations (always using the states showed in Eq. (5.8) as the basis).

---

[14]Many other sets were introduced over the years, particularly in the 1980s and 1990s, including sets containing the three-qubit Deutsch gate and the two-qubit Barenco gate, which are no longer commonly used.

- SWAP:

$$\hat{U}_{\text{SWAP}} = \begin{pmatrix} 1 & 0 & 0 & 0 \\ 0 & 0 & 1 & 0 \\ 0 & 1 & 0 & 0 \\ 0 & 0 & 0 & 1 \end{pmatrix}$$

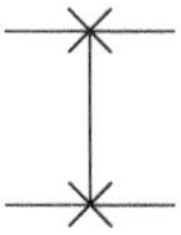

- $\sqrt{\text{SWAP}}$:

$$\hat{U}_{\sqrt{\text{SWAP}}} = \begin{pmatrix} 1 & 0 & 0 & 0 \\ 0 & (1+i)/2 & (1-i)/2 & 0 \\ 0 & (1-i)/2 & (1+i)/2 & 0 \\ 0 & 0 & 0 & 1 \end{pmatrix}$$

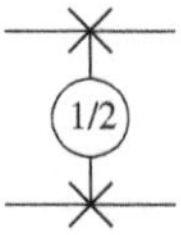

- CSIGN (a.k.a. CTRL-Z):

$$\hat{U}_{\text{CTRL-Z}} = \begin{pmatrix} 1 & 0 & 0 & 0 \\ 0 & 1 & 0 & 0 \\ 0 & 0 & 1 & 0 \\ 0 & 0 & 0 & -1 \end{pmatrix}$$

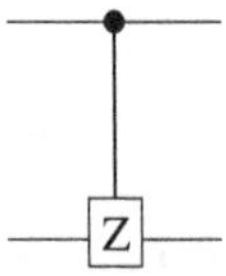

The latter is especially important to optical quantum computing because it is relatively simple to implement in that setting and it is possible to generate other two-qubit gates from it. For instance, consider the circuit equivalence shown in Fig. 5.13.

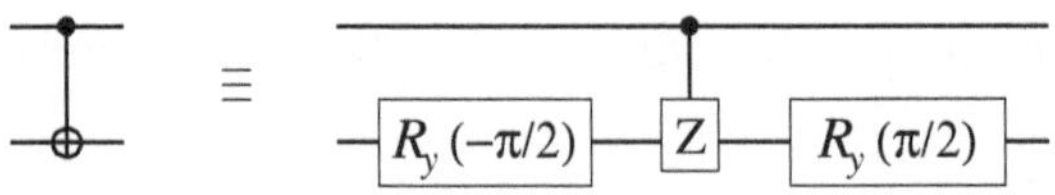

Figure 5.13 Decomposition of a CNOT gate in terms of rotations and a CTRL-Z gate.

## 5.12 REFERENCES AND FURTHER READING

1. Mermin, N. D. 2007. *Quantum Computer Science*. Cambridge Univ. Press. Chapter 1.

2. Nielsen M. A. and I. L. Chuang. 2000. *Quantum Computation and Quantum Information.* Cambridge Univ. Press. Sections 4.2-4.6.

3. Kaye, Ph., R. Laflamme and M. Mosca. 2007. *An Introduction to Quantum Computing.* Oxford Univ. Press. Sections 1.3-1.5 and 4.1-4.5

4. Williams, C. P. 2011. *Explorations in Quantum Computing.* Springer-Verlag. Chapters 2 and 3.

5. Kitaev, A. Yu., A. H. Shen, and M. N. Vyalyi. 2002. *Classical and Quantum Computation.* American Mathematical Society. Section 8.3.

## 5.13 EXERCISES AND PROBLEMS

1. Build a permutation matrix that represents the Toffoli gate. Do not forget to define the vector representation of each possible classical input.

2. Build all possible truth tables for gates in the Toffoli class, namely, for three-bit gates where two bits control a NOT applied on a third (target) bit. Write Boolean functions for each output line of these gates. *Hint:* when there are two control bits, there are $2^2$ different ways for these bits to control the target bit.

3. Show that for any one of the three Pauli gates $\hat{G} = \hat{X}, \hat{Y}, \hat{Z}$,

$$e^{-i\theta\hat{G}/2} = \cos(\theta/2)\,\hat{I} - i\,\sin(\theta/2)\,\hat{G}.$$

*Hint:* use the Taylor expansion mentioned in Ch. 4.

4. Show that a phase gate is equivalent to an $R_z$ gate up to an overall phase factor.

5. Identify the following one-qubit states in the Bloch sphere (use drawings and identify angles):

   (a) $|\psi\rangle = \frac{1}{\sqrt{5}}(i|0\rangle - 2|1\rangle)$

   (b) $|\psi\rangle = e^{i\pi/4}|0\rangle$

   (c) $|\psi\rangle = \frac{i}{\sqrt{2}}(|0\rangle - |1\rangle)$

   (d) $|\psi\rangle = \frac{1}{\sqrt{2}}(|0\rangle + |1\rangle)$

(e) $|\psi\rangle = \frac{i}{2}|0\rangle - \frac{\sqrt{3}}{2}|1\rangle$.

6. Find the change in angles $\theta$ and $\phi$ of a Bloch sphere state when the following gates act on it: (a) $\hat{R}_x(\alpha)$, (b) $\hat{R}_y(\alpha)$, (c) $\hat{R}_z(\alpha)$, (d) $\hat{H}$, (e) $\hat{S}$, and (f) $\hat{T}$.

7. Show that a SWAP gate, when applied to a generic product state of two qubits, always yields another product state. Show that this is not the case for a CNOT gate.

8. Consider a two-qubit state

$$|\Psi\rangle = \frac{1}{\sqrt{2}}\left[ |0\rangle_a|1\rangle_b + |1\rangle_a|0\rangle_b \right].$$

Show that it is an entangled state. *Hint*: prove that there are no possible values of $\alpha_a$, $\beta_a$, $\alpha_b$, and $\beta_b$ such that the state can be reduced to the product

$$|\Psi\rangle = (\alpha_a|0\rangle_a + \beta_a|1\rangle_a) \otimes (\alpha_b|0\rangle_b + \beta_b|1\rangle_b).$$

9. Consider a system of two qubits, $a$ and $b$.

   (a) Show how one can use Hadamard gates to build the following two-qubit basis states from classical ones:

   $$\left\{ |0\rangle_a \otimes \left( \frac{|0\rangle_b + |1\rangle_b}{\sqrt{2}} \right), \ |0\rangle_a \otimes \left( \frac{|0\rangle_b - |1\rangle_b}{\sqrt{2}} \right), \right.$$
   $$\left. |1\rangle_a \otimes \left( \frac{|0\rangle_b + |1\rangle_b}{\sqrt{2}} \right), \ |1\rangle_a \otimes \left( \frac{|0\rangle_b - |1\rangle_b}{\sqrt{2}} \right) \right\}.$$

   (b) Using the Dirac notation, describe the effect of a CNOT gate on these basis states. First, consider qubit $a$ as the control and $b$ as the target, and then vice versa.

10. Consider the two-qubit operator

$$\hat{\Omega} = \frac{1}{2}\hat{I} \otimes \hat{I} + \frac{1}{2}\hat{Z} \otimes \hat{I} + \frac{1}{2}\hat{I} \otimes \hat{X} - \frac{1}{2}\hat{Z} \otimes \hat{X}.$$

   (a) Find a matrix representation for $\hat{\Omega}$. *Hint*: start by defining your computational basis vectors in the two-qubit Hilbert space.

(b) Is the operator $\hat{\Omega}$ unitary? Is it Hermitian? Justify your answers.

(c) Design a quantum circuit (with as many gates as needed) that implements this operator. *Hint*: it is a very simple circuit!

11. Construct a matrix representation for the two-qubit circuit shown in Fig. 5.14. *Hint*: start by defining the two-qubit computational basis states and the action of the gates on these states.

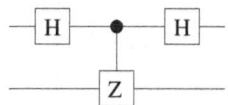

Figure 5.14 Diagram for problem 11.

12. Consider the three-qubit circuits in Fig. 5.15.

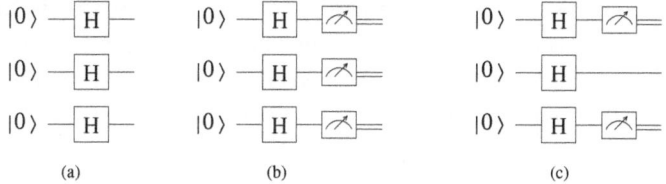

Figure 5.15 Diagrams for problem 12.

(a) Let $|000\rangle$ be the input state. Find the output state for the circuit in panel (a) of Fig. 5.15.

(b) Find the probability that a measurement in the computational basis, as shown in panel (b) of Fig. 5.15, obtains the state $|010\rangle$. *Hint*: define a projector operator for the state $|010\rangle$ and use it to compute the probability amplitude.

(c) Find the probability that a measurement in the computational basis, as shown in panel (c) of Fig. 5.15, obtains the value 1 for the bottom qubit. *Hint*: define a suitable project operator that only acts on the bottom qubit to produce the desired output value for that qubit.

13. Show that the two elementary quantum circuits shown in Fig. 5.16 are equivalent. *Hint*: you can either use unitary matrices or states from a complete basis.

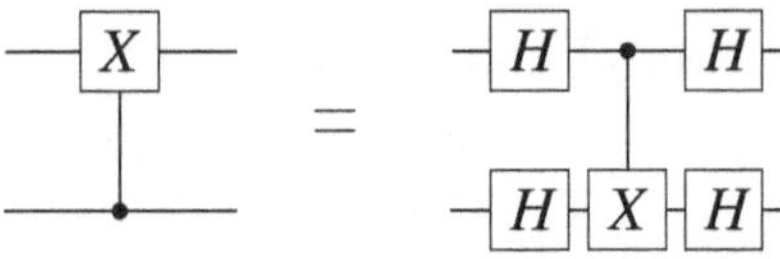

Figure 5.16 Diagram for problem 13.

14. Consider the two-bit function $f(x)$ defined by the truth table

| $x$ | $f(x)$ |
|-----|--------|
| 00 | 01 |
| 01 | 10 |
| 10 | 10 |
| 11 | 01 |

(a) Construct a classical gate (reversible or irreversible), com-posed of Boolean gates, that evaluates this function.

(b) Now construct a quantum gate that evaluates this function. *Hint*: you can build it using classical elementary gates but they need to be reversible. You will need two ancilla qubits; check the approach used in Deutsch's algorithm to turn a Boolean function into a unitary gate.

(c) Obtain a unitary matrix $\hat{U}_f$ that implements the quantum gate. *Hint*: it is a $16 \times 16$ matrix; it can be broken down into sixteen blocks, with each block being a $4 \times 4$ matrix. But only four of the 16 blocks are non trivial.

(d) Using this matrix, find the output state for the input state

$$|\Psi\rangle = \frac{1}{2}(|00\rangle - |01\rangle + |10\rangle - |11\rangle).$$

(e) Repeat the calculation in item d) for the input state $|\Psi\rangle = \hat{H} \otimes \hat{H}|00\rangle$, where $\hat{H}$ denotes a Hadamard gate (i.e., a Hadamard gate is applied to each qubit).

15. Starting from a classical product state, design a quantum circuit to build a superposition involving all possible computational basis states of three qubits. *Hint*: you need at least three gates.

16. Consider the two-qubit state

$$|\Psi\rangle = \frac{1}{\sqrt{3}}(|10\rangle + |01\rangle - |00\rangle).$$

Is this an entangled state or a product state? Justify your answer.

17. Consider the three-qubit state

$$|\Psi\rangle = \frac{1}{\sqrt{6}}(|110\rangle + |010\rangle - 2i|101\rangle).$$

Is this an entangled state or a product state? Justify your answer.

18. Show that the $\hat{H}$ and $\hat{T}$ gates satisfy the conditions of the Solovay-Kitaev theorem.

19. Design quantum circuits to create the following states from classical product states:

   (a) Bell state

$$|\psi\rangle = \frac{1}{\sqrt{2}}(|00\rangle + |11\rangle)$$

   (b) GHZ state

$$|\psi\rangle = \frac{1}{\sqrt{2}}(|000\rangle + |111\rangle).$$

# CHAPTER 6

# Basic Quantum Algorithms

Can an algorithm that explores quantum superposition and entanglement outperform its classical counterpart?

It was only in 1985, long after the birth of quantum mechanics, that an affirmative answer was established. While the concept of a quantum computing machine had been introduced earlier (circa 1982 by Richard Feynman), it was David Deutsch in 1985 who first showed a simple example where a quantum algorithm could beat any classical one.[1]

Deutsch's example is not particularly useful for practical applications, but it does show that superposition and entanglement can be used to speed up a calculation. Using quantum computing jargon, we can say that the algorithm invented by Deutsch illustrates the concepts of quantum parallelism and quantum interference and how they can be combined to speed up the solution of a mathematical problem.

## 6.1 DEUTSCH'S ALGORITHM

What problem does Deutsch's algorithm solve? It determines if a one-bit function is balanced: let $f : \{0, 1\} \rightarrow \{0, 1\}$ denote a function that takes a binary variable into another:

- $f$ is balanced if $f(0) \neq f(1)$;

- $f$ is not balanced if $f(0) = f(1)$ (i.e., it is a constant in this particular one-bit case).

---

[1]Deutsch, D. 1985. *Quantum Theory, the Church-Turing Principle and the Universal Quantum Computer*. Proc. R. Soc. Lond. A 400:97–117.

Imagine that $f$ is implemented by a reversible circuit inside a black box. We cannot tell how $f$ works internally, therefore cannot predict in advance if $f$ is balanced unless we test it. How many queries to the black box do we have to make?

- Classically, the answer is two: query $f(0)$ and $f(1)$ and then compare the results.

- Quantum mechanically, the answer is one! How is that possible?!

Deutsch's algorithm finds $f(0) \oplus f(1)$ without querying $f(0)$ and $f(1)$ separately; it queries the black box only once. From knowing $z = f(0) \oplus f(1)$ we can immediately tell if $f$ is balanced or not: a balanced $f$ yields $z = 0$ while a constant one yields $z = 1$. Since $f$ is entirely made out of reversible gates (although classical), it can be realized it with quantum gates without any constraints.

Let $\hat{U}_f$ represent the two-qubit unitary operator that implements the following control gate:

$$\hat{U}_f |x\rangle |y\rangle = |x\rangle |y \oplus f(x)\rangle.$$

Namely, $\hat{U}_f$ is such that:

- if $x = 0$ (1st qubit) and $y = 0$ (2nd qubit), get $|f(0)\rangle$ on the 2nd qubit;

- if $x = 0$ and $y = 1$, get $|1 \oplus f(0)\rangle = |\overline{f(0)}\rangle$ on the 2nd qubit;

- if $x = 1$ and $y = 0$, get $|f(1)\rangle$ on the 2nd qubit;

- if $x = 1$ and $y = 1$, get $|1 \oplus f(1)\rangle = |\overline{f(1)}\rangle$ on the 2nd qubit.

(Recall that a bar on a binary variable means its negation.) What happens when we start with $|x\rangle = \frac{1}{\sqrt{2}}(|0\rangle + |1\rangle)$ and $|y\rangle = |0\rangle$ on input? Let us work it out:

$$\hat{U}_f \left( \frac{|0\rangle + |1\rangle}{\sqrt{2}} \right) |0\rangle = \frac{1}{\sqrt{2}} \hat{U}_f |0\rangle |0\rangle + \frac{1}{\sqrt{2}} \hat{U}_f |1\rangle |0\rangle$$

$$= \frac{1}{\sqrt{2}} |0\rangle |f(0)\rangle + \frac{1}{\sqrt{2}} |1\rangle |f(1)\rangle.$$

The output state is a superposition of both $f(0)$ and $f(1)$ on the second qubit. In other words, both computations have happened! How do we extract $f(0) \oplus f(1)$? Use a superposition state for the other qubit as well. To do so, let us start instead with the initial state

$$|\psi_0\rangle = \left(\frac{|0\rangle + |1\rangle}{\sqrt{2}}\right)\left(\frac{|0\rangle - |1\rangle}{\sqrt{2}}\right).$$

Again, let us work it out:

$$
\begin{aligned}
|\psi_1\rangle &= \hat{U}_f|\psi_0\rangle \\
&= \hat{U}_f\left(\frac{1}{2}|00\rangle - \frac{1}{2}|01\rangle + \frac{1}{2}|10\rangle - \frac{1}{2}|11\rangle\right) \\
&= \frac{1}{2}(\hat{U}_f|00\rangle - \hat{U}_f|01\rangle + \hat{U}_f|10\rangle - \hat{U}_f|11\rangle) \\
&= \frac{1}{2}\left[|0\rangle|0 \oplus f(0)\rangle - |0\rangle|1 \oplus f(0)\rangle + |1\rangle|0 \oplus f(1)\rangle - |1\rangle|1 \oplus f(1)\rangle\right] \\
&= \frac{1}{2}\left[|0\rangle|f(0)\rangle - |0\rangle|1 \oplus f(0)\rangle + |1\rangle|f(1)\rangle - |1\rangle|1 \oplus f(1)\rangle\right] \\
&= \frac{1}{2}\{|0\rangle(|f(0)\rangle - |1 \oplus f(0)\rangle) + |1\rangle(|f(1)\rangle - |1 \oplus f(1)\rangle)\}.
\end{aligned}
$$

Now we can make use of the following relation:

$$
\begin{aligned}
|f(x)\rangle - |1 \oplus f(x)\rangle &= \begin{cases} |0\rangle - |1\rangle & \text{if } f(x) = 0 \\ |1\rangle - |0\rangle & \text{if } f(x) = 1 \end{cases} \\
&= (-1)^{f(x)}(|0\rangle - |1\rangle).
\end{aligned}
$$

Therefore,

$$
\begin{aligned}
|\psi_1\rangle &= \frac{1}{2}\{(-1)^{f(0)}|0\rangle\left[|0\rangle - |1\rangle\right] + (-1)^{f(1)}|1\rangle\left[|0\rangle - |1\rangle\right]\} \\
&= \frac{1}{2}\left[(-1)^{f(0)}|0\rangle + (-1)^{f(1)}|1\rangle\right](|0\rangle - |1\rangle) \\
&= (-1)^{f(0)}\left[\frac{|0\rangle + (-1)^{f(0) \oplus f(1)}|1\rangle}{\sqrt{2}}\right]\frac{(|0\rangle - |1\rangle)}{\sqrt{2}}.
\end{aligned}
$$

Now, apply a Hadamard gate on the first qubit:

$$|\psi_2\rangle = \hat{H}_{\text{qubit 1}}|\psi_1\rangle.$$

- If $f(0) \oplus f(1) = 1$ (balanced function), then

$$|\psi_2\rangle = (-1)^{f(0)}|1\rangle\left(\frac{|0\rangle - |1\rangle}{\sqrt{2}}\right);$$

- if $f(0) \oplus f(1) = 0$ (constant function), then

$$|\psi_2\rangle = (-1)^{f(0)}|0\rangle \left(\frac{|0\rangle - |1\rangle}{\sqrt{2}}\right).$$

By measuring the state of the first qubit in the computational basis after applying the Hadamard gate, we can with certainty find out if the function $f$ is balanced or not! The state of the qubit tells us that:

$$1\text{st qubit} = |1\rangle \quad \Rightarrow \quad \text{balanced}$$
$$1\text{st qubit} = |0\rangle \quad \Rightarrow \quad \text{constant}.$$

A quantum circuit that implements this algorithm is shown in Fig. 6.1.

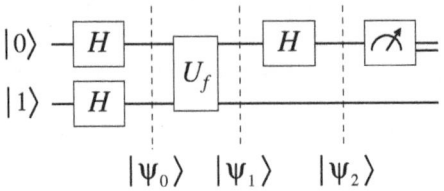

**Figure 6.1** Quantum circuit for Deutsch's algorithm.

In Fig. 6.1, we assume that there is a reversible-circuit implementation of the operator $\hat{U}_f$. The dashed lines indicate the stage where the state vectors were computed in the derivation developed above.

Deutsch's algorithm determines whether $f$ is balanced or not by a single query to the classical black-box circuit that computes $f$ (which is assumed to be inside $\hat{U}_f$). It does so at the expense of an extra (ancilla) qubit and a few single-qubit gates.

Notice that in the process of implementing this algorithm we employed superposition and entanglement. Can you tell on which stage each one of these properties were used? In this particular case, quantum, interference led us to the desired answer with 100% certainty. We will see later that this is not always the case with quantum algorithms.

## 6.2 DEUTSCH-JOZSA ALGORITHM

Deutsch's algorithm can be generalized to find whether an $n$-bit function $f : \{0,1\}^n \rightarrow \{0,1\}$ is balanced (i.e., 0 for half of the possible $2^n$ entries and 1 for the other half) or a constant (0 or 1 for all $2^n$ entries), provided that we know that it can only be one or the other. The resulting algorithm is called Deutsch-Jozsa,[2, 3] which requires $n + 1$ qubits and an operator that acts on this larger space according to the rule,

$$\hat{U}_f|x_1 \cdots x_n\rangle|y\rangle = |x_1 \cdots x_n\rangle|y \oplus f(x_1, \ldots, x_n)\rangle.$$

The circuit that implements this algorithm is shown in Fig. 6.2

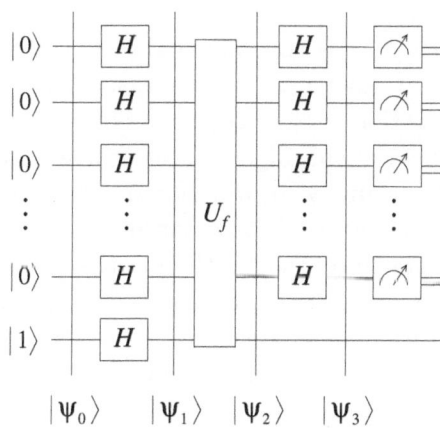

Figure 6.2 Quantum circuit for the Deutsch-Jozsa algorithm.

Notice the $(n+1)$-qubit $\hat{U}_f$ gate and the $n$ measurements. We assume that a quantum implementation of $\hat{U}_f$ requires a computational effort that scales polynomially with the number of qubits.

Following the same approach used for the Deutsch algorithm, we can establish the evolution of the quantum state at the stages indicated by the dashed lines in Fig. 6.2:

$$|\psi_0\rangle = \left[\prod_{k=1}^{n}|0\rangle_k\right] \otimes |1\rangle_{n+1},$$

$$|\psi_1\rangle = \frac{1}{2^{n/2}} \left[\sum_{x=0}^{2^n-1}|x\rangle\right] \otimes \frac{1}{\sqrt{2}}\left(|0\rangle - |1\rangle\right),$$

$$|\psi_2\rangle = \frac{1}{2^{n/2}} \left[\sum_{x=0}^{2^n-1}(-1)^{f(x)}|x\rangle\right] \otimes \frac{1}{\sqrt{2}}\left(|0\rangle - |1\rangle\right), \tag{6.1}$$

[2]Deutsch, D. and R. Jozsa. 1992. *Rapid solutions of problems by quantum computation.* Proc. R. Soc. Lond. A 439: 553-558

[3]Cleve, R., A. Ekert, C. Macchiavello, and M. Mosca. 1998. *Quantum algorithms revisited.* Proc. R. Soc. Lond. A 454: 339-354

and

$$
\begin{aligned}
|\psi_3\rangle &= \sum_{x=0}^{2^n-1} \left[ \frac{1}{2^n} \sum_{x'=0}^{2^n-1} (-1)^{f(x)}(-1)^{x \cdot x'} |x'\rangle \right] \otimes \frac{1}{\sqrt{2}} (|0\rangle - |1\rangle) \\
&= \sum_{x=0}^{2^n-1} p(x) |x\rangle \otimes \frac{1}{\sqrt{2}} (|0\rangle - |1\rangle) , \quad\quad (6.2)
\end{aligned}
$$

where $p(0) = 1$ when $f(x)$ is constant and $p(0) = 0$ when $f(x)$ is balanced. Therefore, if at least one qubit measurement yields 1, we can conclude that the function is balanced.

The amazing aspect of this algorithm is that it allows for the determination (with 100% accuracy) of whether an $n$-bit function is balanced or constant through *a single query* to the black box that computes $f$. No classical algorithm can do that in the most general case![4] Of course, the price to pay is the use of $n$ ancilla qubits. But this may not be a high price given that we may have avoided $2^n$ classical computations. So, overall, we replaced an exponential number of classical computations of $f$ by a polynomial number of quantum operations (i.e., proportional to some power of $n$) on the $n + 1$ qubits. More precisely, we replaced the $O(2^n)$ worst-case-scenario computational cost of the classical algorithm with a quantum one that costs only $O(n^\alpha)$, for some $\alpha > 0$. For large enough $n$, $O(n^\alpha) > O(2^n)$, and by a lot.

Quantum parallelism is achieved in this case by the battery of Hadamard gates at the first layer of the quantum circuit, which puts the $(n + 1)$ qubits into a state comprising a superposition of all $2^{n+1}$ computational basis states and with equal amplitudes.

The second battery of Hadamard gates after the control gate steers the many-qubit state toward another superposition that reveals the answer to you query, and no more. Quantum interference allows us to filter the bit of information we are seeking without having to run the circuit more than once. We only access the information we need!

The Deutsch-Jozsa algorithm shows the exponential advantage of a quantum computer over a classical one for a deterministic task such as finding if a binary function is balanced or not. However, there is a caveat. There are probabilistic algorithms that can, in principle, solve the Deutsch-Jozsa problem with probability at most $1/3$ with only two

---

[4]Someone using a classical algorithm that queries $f(x_1, \ldots, x_n)$ may be lucky and find out, after only two queries, that $f$ is not a constant (and therefore must be balanced). But in the worst-case scenario, they may have to make as many as $2^{n-1}$ queries to arrive at a conclusion.

queries. The probability of error can be further reduced to $1/2^n$ with $n+1$ queries. So, the quantum advantage of the Deutsch-Jozsa algorithm is less impressive against a classical probabilistic algorithm.

## 6.3 SIMON'S ALGORITHM

There is a problem solvable efficiently by a quantum circuit where even a classical probabilistic algorithm is exponentially worse. It is the Simon's problem:[5]

Consider the function $f : \{0,1\}^n \to Q$, where $Q$ is some finite set of bit states. Suppose that a bit string $s_1 s_2 \cdots s_n$ exists such that $f(x_1 x_2 \cdots x_n) = f(y_1 y_2 \cdots y_n)$ if and only if $x_1 x_2 \cdots x_n = y_1 y_2 \cdots y_n$ or $x_1 x_2 \cdots x_n = s_1 \oplus y_1 s_2 \oplus y_2 \cdots s_n \oplus y_n$. Find the string $s_1 s_2 \cdots s_n$ by making queries to $f$.

It turns out that probabilistic classical algorithms can solve this problem at least 2/3 of the time, but only after evaluating $f$ $O(2^{n/3})$ times! However, there is a quantum algorithm invented by Daniel Simon in 1997 that queries $f$ less than $O(n)$ times and *always* finds a solution. It requires $O(n^3)$ gates and $O(n)$ ancilla bits.

One can do even better and use fewer queries by allowing probabilistic success of the quantum algorithm.

## 6.4 REFERENCES AND FURTHER READING

1. Mermin, N. D. 2007. *Quantum Computer Science*. Cambridge Univ. Press. Chapter 2.

2. Nielsen M. A. and I. L. Chuang. 2000. *Quantum Computation and Quantum Information*. Cambridge Univ. Press. Section 1.4.

3. Kaye, Ph., R. Laflamme and M. Mosca. 2007. *An Introduction to Quantum Computing*. Oxford Univ. Press. Chapter 6.

## 6.5 EXERCISES AND PROBLEMS

1. Obtain Eq. (6.2) from Eq. (6.1) by applying a Hadamard to each one of the first $n$ qubits. Obtain an expression for $p(x)$. Show that $p(0) = \pm 1$ corresponds to a constant function and $p(0) = 0$ to a balanced function.

[5]Simon, D. R. 1997. *On the power of quantum computation*. SIAM J. Comput. 26: 1474-1483

2. Consider $x \in \{0, 1\}^n$, i.e., an $n$-bit string. Let $y = x_1 \oplus x_2 \oplus \cdots \oplus x_n$ define the parity of the bit string. Classically, to determine $y$, we would need to query the bit string $n$ times (to perform XOR operations). Show that using the Deutsch-Jozsa algorithm, we can reduce the number of queries to $n/2$.

# CHAPTER 7

# Quantum Information: Limits and Possibilities

Even though quantum mechanics allows a state to pack a lot of information (via arbitrary superposition of basis states), not all information contained in a quantum state is retrievable from measurements. This seems to be a limitation at first, but can also offer some advantages over classical ways to store and transmit information.

## 7.1 NO-CLONING THEOREM

To understand why one cannot retrieve all information stored in a quantum state, consider the following situation.

Alice has a qubit in the state

$$|\psi\rangle = a|0\rangle + b|1\rangle,$$

with $|a|^2 + |b|^2 = 1$. She wants to build her own copy of it and send it to Bob. For this purpose, she performs a measurement of the operator

$$\hat{O} = |1\rangle\langle 1|$$

on the state $|\psi\rangle$. Notice that this operator provides some information about the amount of $|1\rangle$ present in $|\psi\rangle$.

What values can she obtain? Answer: 0 and 1. The frequency of occurrence of each outcome depends on the amplitudes $a$ and $b$. She needs to repeat the experiment many times to try to pin down $a$ and $b$.

What are the probabilities of each one of the two possible outcomes? Answer: $p(0) = |a|^2$ and $p(1) = |b|^2 = 1 - |a|^2$.

Suppose that $a = 1/2$. Alice has been given 1000 identical samples of the state $|\psi\rangle$ (but still wants to build one herself). After repeating the same measurement of $\hat{Q}$ for each sample, she would be able to tell that $|a| \approx 1/2$. She may even assume without loss of generality that $a \approx 1/2$.

After obtaining $a$, can she determine $b$ accurately? Answer: not really. She can certainly estimate that $|b| \approx \sqrt{3}/2$ with a relatively small error (because of the large number of samples), but she cannot distinguish $b \approx \sqrt{3}/2$ from $b \approx i\sqrt{3}/2$, for instance. She cannot determine the phase of $b$, which is a nontrivial relative phase since she chose $a$ to be real. Because of that, she cannot create another state exactly like $|\psi\rangle$. Even if someone gives her another 1000 copies, she would not be able to build her own $|\psi\rangle$.

This result can be established more rigorously via the "no-cloning theorem":[1, 2]

*There is no unitary operator $\hat{U}$ that satisfies the relation $\hat{U}|\psi\rangle|\phi\rangle = |\psi\rangle|\psi\rangle$ for arbitrary $|\psi\rangle$.*

In other words, there is no way to evolve a quantum system to exactly replicate the content of another quantum system.

The proof is rather simple and works by contradiction. Suppose that $\hat{U}$ existed. Then,

$$\hat{U}|\Omega\rangle|\phi\rangle = |\Omega\rangle|\Omega\rangle$$

must be true for any $|\Omega\rangle \neq |\psi\rangle$ as well. Then,

$$\langle\Omega|\langle\Omega| = \langle\Omega|\langle\phi|\hat{U}^\dagger,$$

which we can use to write

$$\langle\Omega|\langle\Omega| \cdot |\psi\rangle|\psi\rangle = \langle\phi|\langle\Omega|\hat{U}^\dagger\hat{U}|\psi\rangle|\phi\rangle.$$

Since $\hat{U} = \hat{U}^\dagger$, then

$$\langle\Omega|\langle\Omega| \cdot |\psi\rangle|\psi\rangle = \langle\phi|\langle\Omega| \cdot |\psi\rangle|\phi\rangle.$$
$$\downarrow \qquad\qquad \downarrow$$
$$\langle\Omega|\psi\rangle^2 = \langle\Omega|\psi\rangle$$

The latter can only be satisfied if $\langle\Omega|\psi\rangle = 0$ or $\langle\Omega|\psi\rangle = 1$. But $|\Omega\rangle$ is supposed to be arbitrary and these conditions are very restrictive.

---

[1] Wootters, W. and W. Zurek. 1982. *A single quantum cannot be cloned.* Nature 299: 802-803

[2] Dieks, D. 1982. *Communication by EPR devices.* Phys. Lett. A 92:271-272

Therefore, we can only conclude that the assumption that $\hat{U}$ exists is untenable.

Because it is impossible to clone quantum states, they provide a very secure way to store information.[3] Yet, one can transmit a quantum state even without being able to clone or fully determine it via measurements! The way to do it is via *teleportation*.

## 7.2 QUANTUM TELEPORTATION

Not being able to fully retrieve information about the state of a qubit does not prevent one from reconstituting that state elsewhere!

It turns out that Alice can send a qubit to Bob using a classical channel even without knowing exactly the state of the qubit she has, provided that they share a two-qubit entangled state.[4]

To understand how this works, we need to introduce a family of two-qubit entangled states called Bell states (named after John Bell):

$$|B_{00}\rangle = \frac{1}{\sqrt{2}}(|00\rangle + |11\rangle)$$

$$|B_{01}\rangle = \frac{1}{\sqrt{2}}(|01\rangle + |10\rangle)$$

$$|B_{10}\rangle = \frac{1}{\sqrt{2}}(|00\rangle - |11\rangle)$$

$$|B_{11}\rangle = \frac{1}{\sqrt{2}}(|01\rangle - |10\rangle).$$

Suppose that Alice, in addition to having a qubit with state $|\psi\rangle$, she and Bob share a state $|B_{00}\rangle$. When we say share, we mean that Alice has in her possession one of the qubits making up the state $|B_{00}\rangle$ while Bob has the other one.[5] Call

$$|\psi\rangle = \alpha|0\rangle + \beta|1\rangle$$

the state Alice wants to send to Bob, with $|\alpha|^2 + |\beta|^2 = 1$. Instead of physically transporting this qubit to Bob, she can transmit the information encoded in $|\psi\rangle$ to a qubit that is already in Bob's possession, namely, to Bob's qubit that is entangled to Alice's other qubit.

---

[3]One may even say it is too secure since retrieving the complete information stored therein is impossible!

[4]Bennett, C. H., G. Brassard, C. Crepeau, R. Jozsa, A. Peres, W. K. Wootters. 1993. *Teleporting an unknown quantum state vial dual classical and Einstein-Podolsky-Rosen channels.* Phys. Rev. Lett. 70: 1895-1899

[5]This means that sometime in the past Alice and Bob were able to let their qubits interact long enough to become entangled.

Notice that there are three qubits in this story: Alice has two and Bob has one (which is entangled with one of Alice's qubits, but not with the other).

The initial state of the 3-qubit system is

$$|\psi\rangle \otimes |B_{00}\rangle = (\alpha|0\rangle + \beta|1\rangle) \otimes \frac{1}{\sqrt{2}}(|00\rangle + |11\rangle).$$

- *Interlude*: how can this state be obtained? One way is to apply the circuit of Fig. 7.1. Imagine that Alice has a qubit (e.g., a photon) which she initializes to 0, passes through a Hadamard gate, and then sends to Bob via a quantum channel (e.g., a noiseless quantum fiber that preserves quantum coherence) to act as a control of a CNOT gate he applies to a target qubit that he had initialized to 0.

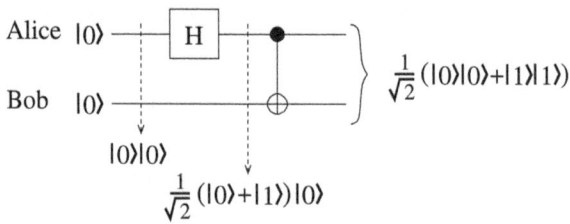

Figure 7.1  Circuit to prepare Bell's $|B_{00}\rangle$ state.

Let us look more closely at the $|\psi\rangle \otimes |B_{00}\rangle$ state. Expanding the tensor product, we arrive at

$$|\psi\rangle \otimes |B_{00}\rangle = \frac{1}{\sqrt{2}}(\alpha|000\rangle + \alpha|011\rangle + \beta|100\rangle + \beta|111\rangle).$$

We can rewrite each term on the r.h.s. of this expression in a way that reveals a little more about their content.[6] For instance,

$$
\begin{aligned}
\alpha|000\rangle &= \frac{1}{2}[\alpha|000\rangle + \alpha|000\rangle] \\
&= \frac{1}{2}[\alpha|000\rangle + \beta|001\rangle + \alpha|000\rangle - \beta|001\rangle] \\
&= \frac{1}{2}[|00\rangle \otimes (\alpha|0\rangle + \beta|1\rangle) + |00\rangle(\alpha|0\rangle - \beta|1\rangle)] \\
&= \frac{1}{2}\left[|00\rangle \otimes |\psi\rangle + |00\rangle \otimes \hat{Z}|\psi\rangle\right].
\end{aligned}
$$

---

[6]One may ask how could one have come up with such an idea? I ask myself the same question, but one may suspect that physical intuition led the inventors to teleportation. Then they had to figure out a mathematical way to derive their result.

Notice that we rearranged the terms in such a way that now $|\psi\rangle$ shows up on the third qubit (i.e., on Bob's). Similarly,

$$\beta|111\rangle = \frac{1}{2}[\beta|111\rangle + \alpha|110\rangle + \beta|111\rangle - \alpha|110\rangle]$$

$$= \frac{1}{2}\left[|11\rangle \otimes |\psi\rangle - |11\rangle \otimes \hat{Z}|\psi\rangle\right]$$

$$\alpha|011\rangle = \frac{1}{2}[\alpha|011\rangle + \beta|010\rangle + \alpha|011\rangle - \beta|010\rangle]$$

$$= \frac{1}{2}\left[|01\rangle \otimes \hat{X}|\psi\rangle + |01\rangle \otimes \hat{X}\hat{Z}|\psi\rangle\right]$$

$$\beta|100\rangle = \frac{1}{2}\left[\beta|100\rangle + \frac{1}{2}\alpha|101\rangle + \beta|100\rangle - \alpha|101\rangle\right]$$

$$= \frac{1}{2}\left[|10\rangle \otimes \hat{X}|\psi\rangle - |10\rangle \otimes \hat{X}\hat{Z}|\psi\rangle\right].$$

Putting it all together, we find

$$|\psi\rangle \otimes |B_{00}\rangle = \frac{1}{2}\left[|B_{00}\rangle \otimes |\psi\rangle + |B_{01}\rangle \otimes \hat{X}|\psi\rangle + |B_{10}\rangle \otimes \hat{Z}|\psi\rangle\right.$$
$$\left. + |B_{11}\rangle \otimes \hat{X}\hat{Z}|\psi\rangle\right].$$

This is interesting because the state $|\psi\rangle$ appears in various forms on Bob's qubit. If Alice measures her qubits, she can tell Bob which operators he needs to apply on his qubit to steer it toward the state $|\psi\rangle$:

- If Alice measures $B_{00}$, she tells Bob that he got $|\psi\rangle$ already; no operations are required.

- If Alice measures $B_{01}$, she tells Bob to apply $\hat{X}$ on his qubit to turn it into $|\psi\rangle$.

- If Alice measures $B_{10}$, she tells Bob to apply $\hat{Z}$ instead.

- Finally, if Alice measures $B_{11}$, she tells Bob to first apply $\hat{X}$ and then $\hat{Z}$ on his qubit.

Alice can transmit her instructions to Bob via classical channels (e.g., a telephone line or a text message where only bits are transmitted). The circuit that implements the teleportation protocol is shown in Fig. 7.2.

Because Alice and Bob share a two-qubit entangled state, Alice can teleport an entire qubit state to Bob by only transmitting *two bits* of information. The power of entanglement!

Quantum teleportation has some important applications in quantum circuitry and quantum communications which we will discuss later on.

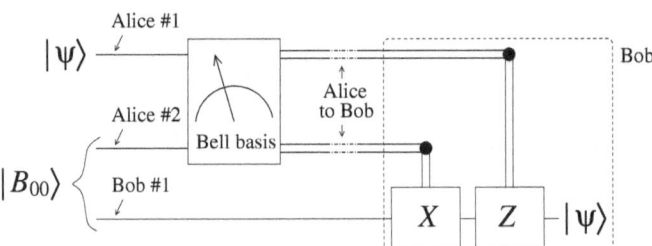

Figure 7.2  Circuit for teleportation.

## 7.3  SUPERDENSE CODING

The inverse situation, namely, sending two classical bits of information in a single qubit, is also possible if the parties share an entangled state and are connected by a quantum channel . This is referred as superdense coding.[7] How does it work?

Support that Alice and Bob share the Bell state

$$|B_{00}\rangle = \frac{1}{\sqrt{2}}(|00\rangle + |11\rangle)$$

(the first qubit is Alice's and the second qubit is Bob's).

- If Alice wants to send "00" to Bob, she does nothing to her qubit (which is equivalent to applying the identity operator).

- If Alice wants to send "01" to Bob, she applies an $\hat{X}$ gate to her qubit.

- To send "10", she applies a $\hat{Z}$ gate.

- To send "11", she first applies an $\hat{X}$ gate and then a $\hat{Z}$ gate.

Once Alices is done with her operations, she sends her qubit to Bob, who measures the two-qubit system (now in his possession) in a Bell basis, namely, $\{|B_{00}\rangle, |B_{01}\rangle, |B_{10}\rangle, |B_{11}\rangle\}$. The result of Bob's measurement encodes the two-bit message that Alice wanted to convey to Bob.

---

[7]Bennett, C. and S. Wiesner. 1992. *Communication via one- and two-particle operators on Einstein-Podolsky-Rosen states.* Phys. Rev. Lett. 69: 2881-2884

In math form:

$$"00": \quad \frac{1}{\sqrt{2}}(|00\rangle + |11\rangle) \quad \begin{array}{c} \hat{I} \otimes \hat{I} \\ = \end{array} \quad \frac{1}{\sqrt{2}}(|00\rangle + |11\rangle) = |B_{00}\rangle$$

$$"01": \quad \frac{1}{\sqrt{2}}(|00\rangle + |11\rangle) \quad \begin{array}{c} \hat{X} \otimes \hat{I} \\ = \end{array} \quad \frac{1}{\sqrt{2}}(|10\rangle + |01\rangle) = |B_{01}\rangle$$

$$"10": \quad \frac{1}{\sqrt{2}}(|00\rangle + |11\rangle) \quad \begin{array}{c} \hat{Z} \otimes \hat{I} \\ = \end{array} \quad \frac{1}{\sqrt{2}}(|00\rangle - |11\rangle) = |B_{10}\rangle$$

$$"11": \quad \frac{1}{\sqrt{2}}(|00\rangle + |11\rangle) \quad \begin{array}{c} \hat{Z} \cdot \hat{X} \otimes \hat{I} \\ = \end{array} \quad \frac{1}{\sqrt{2}}(|01\rangle - |10\rangle) = |B_{11}\rangle.$$

The digram in Fig. 7.3 shows a circuit representation of this protocol.

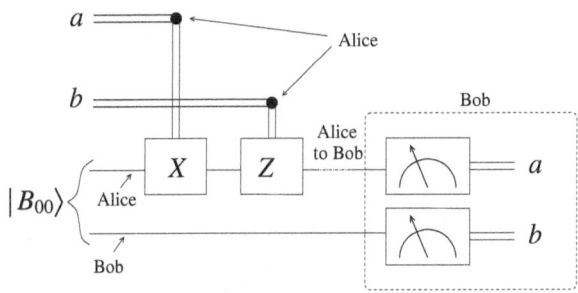

**Figure 7.3** Circuit for superdense coding.

Notice that this form of transmitting information is also secure: if anyone intercepts Alice's qubit and measures it, they will not gain any information, as the outcome will produces either 0 or 1 with equal probability. In fact, the intercepted qubit is in a maximally mixed state.

## 7.4 ENTANGLEMENT SWAPPING

A very useful feature of multi-component quantum mechanical systems is entanglement swapping: given two pairs of entangled systems, by taking one system from each pair and projecting the new pair onto a Bell state, the remaining pair becomes entangled even thought the systems in this pair have not directly interacted.[8] This is a form of quantum teleportation, as one does not need to completely know the state of all systems

---

[8]Zukowski, M., A. Zeilinger, M. A. Horne, and A. Ekert. 1993. *"Event-ready-detectors" Bell experiment via entanglement swapping.* Phys. Rev. Lett. 71:4287-4290

involved. Such an entanglement swapping is the basis of quantum repeaters, which are used to increase the range of transmission channels in a quantum network.

Here is how it works. Let $(A, B)$ and $(C, D)$ be the initial pairs of entangled systems: $A$ is entangled with $B$ and $C$ is entangled with $D$. The initial state vector of the total system can be written as a product:

$$|\Psi_{\text{in}}\rangle = |\psi\rangle_{AB} \otimes |\phi\rangle_{CD}.$$

For simplicity, let us assume that each system consists of a single qubit, in which case we can write the following general expressions for the entangled pair states:

$$|\psi\rangle_{AB} = \alpha_{00}|00\rangle_{AB} + \alpha_{01}|01\rangle_{AB} + \alpha_{10}|10\rangle_{AB} + \alpha_{11}|11\rangle_{AB}$$

and

$$|\phi\rangle_{CD} = \beta_{00}|00\rangle_{CD} + \beta_{01}|01\rangle_{CD} + \beta_{10}|10\rangle_{CD} + \beta_{11}|11\rangle_{CD},$$

where the amplitudes are unknown. Now, let us apply the state projector

$$\hat{P}_{B_{00}} = |B_{00}\rangle_{BC}\,_{BC}\langle B_{00}|,$$

where the Bell state involves systems $B$ and $C$:

$$|B_{00}\rangle_{BC} = \frac{1}{\sqrt{2}}\left(|00\rangle_{BC} + |11\rangle_{BC}\right).$$

Applying the project $\hat{P}_{B_{00}}$ onto the product state $|\Psi_{\text{in}}\rangle$, we obtain (after a bit of algebra) another product,

$$\begin{aligned}
|\Psi_{\text{fin}}\rangle &= \hat{P}_{B_{00}}|\Psi_{\text{in}}\rangle \\
&= |B_{00}\rangle_{BC} \otimes |\omega\rangle_{AD},
\end{aligned}$$

where

$$|\omega\rangle_{AD} = \frac{1}{\sqrt{2}}\left(\gamma_{00}|00\rangle_{AD} + \gamma_{01}|01\rangle_{AD} + \gamma_{10}|10\rangle_{AD} + \gamma_{11}|11\rangle_{AD}\right)$$

and the amplitudes are defined by the relations

$$\begin{aligned}
\gamma_{00} &= \alpha_{00}\beta_{00} + \alpha_{01}\beta_{10} \\
\gamma_{01} &= \alpha_{00}\beta_{01} + \alpha_{01}\beta_{11} \\
\gamma_{10} &= \alpha_{10}\beta_{00} + \alpha_{11}\beta_{10} \\
\gamma_{11} &= \alpha_{10}\beta_{01} + \alpha_{11}\beta_{11}.
\end{aligned}$$

Notice that, after the projection, the systems $A$ and $D$ form an entangled pair. Because the states of the initial pairs $(A, B)$ and $(C, D)$ were unknown (but presumed entangled), the state of the pair $(A, D)$ is also unknown (i.e., its amplitudes are unknown but uniquely defined by the amplitudes of the initial state). The only thing known is the state of the entangled pair $(B, C)$ since it was projected onto a desired Bell state ($B_{00}$ in this example). Notice that since we went from 8 to 4 unknowns, information is not entirely preserved in the process.

## 7.5  BELL'S INEQUALITY

We have not discussed much the foundations of quantum mechanics and its various interpretations. But there is one result related to these topics which deserves some explanation, as it reveals how much quantum mechanics departs from our intuitive understanding of nature. Entire books have been devoted to it but here we will only provide the essentials and follow closely an argument named CHSH after its authors.[9]

Suppose that Charlie prepares pairs of qubits and then sends a qubit of the pair to Alice and another to Bob. Alice and Bob each can measure certain properties of these qubits: $A_1$ and $A_2$ for Alice and $B_1$ and $B_2$ for Bob. Let us assume that the outcome of each one of these possible measurements is $\pm 1$. Each time Alice and Bob receive their qubits, they either measure one or the other property they have access to, picking which property randomly and with equal probability. Therefore, for each qubit pair received by Alice and Bob, there are four possible measurements being performed: $(A_1, B_1)$, $(A_1, B_2)$, $(A_2, B_1)$, and $(A_2, B_2)$. By receiving a large number of qubits, Alice and Bob can build a very accurate estimate of the probability distribution of these four possible joint measurements.

Non-quantum common sense indicates that we can make two hypotheses about this experiment:

- The results of the measurements are predetermined and the only reason why probabilities are being computed is because Alice and Bob do not have access to some hidden variables that deterministically describe the state of the qubits received from Charlie.

- Alice's and Bob's measurements are performed independently of each other, namely, they do not affect each other.

---

[9]Clauser, J. F, M. A. Horne, A. Shimony, and R. A. Holt. 1969. *Proposed experiment to test local hidden-variable theorem.* Phys. Rev. Lett. 23:880-884

Let us carry on and check the consequences of these hypotheses. Consider the quantity

$$Q = A_1 B_1 + A_2 B_1 - A_1 B_2 + A_2 B_2 = A_1 (B_1 - B_2) + A_2 (B_1 + B_2). \quad (7.1)$$

Since each variable in this expression is limited to be $\pm 1$, it is easy to see that $Q = \pm 2$ for each pair of qubits received and measured by Alice and Bob. Therefore, after multiple runs, the following inequality must hold:

$$\left| \overline{A_1 B_1} + \overline{A_2 B_1} - \overline{A_1 B_2} + \overline{A_2 B_2} \right| \leq 2, \quad (7.2)$$

where the overline indicates statistical average. This is a special case of a more general class of inequalities first derived by John Bell.

Let us now check what quantum mechanics can tell us about this situation. Assuming that each time Charlie prepares a qubit pair, it is in the state

$$|\psi\rangle = \frac{1}{\sqrt{2}} \left( |0\rangle_A |1\rangle_B - |1\rangle_A |0\rangle_B \right). \quad (7.3)$$

Moreover, assume that the quantities measured by Alice and Bob are the following:

$$A_1 = Z, \quad A_2 = X, \quad B_1 = (X + Z)/\sqrt{2}, \quad B_2 = (X - Z)/\sqrt{2}. \quad (7.4)$$

Utilizing the state in Eq. (7.2), after a bit of algebra, one finds the following expectation values:

$$\langle A_1 B_1 \rangle = \langle A_2 B_1 \rangle = \langle A_2 B_2 \rangle = -\langle A_1 B_2 \rangle = -\frac{1}{\sqrt{2}}. \quad (7.5)$$

Therefore, according to quantum mechanics,

$$|\langle A_1 B_1 \rangle + \langle A_2 B_1 \rangle - \langle A_1 B_2 \rangle + \langle A_2 B_2 \rangle| = 2\sqrt{2}, \quad (7.6)$$

which does not satisfy the inequality in Eq. (7.3)! This result has been verified experimentally multiple times and with very high accuracy in the past 50 years. It clearly demonstrates that common-sense assumptions break down in quantum mechanics. Either the hidden-variable or the locality hypotheses (or both!) do not hold. It led John Bell to the following theorem (here rephrased from the original for clarity):[10]

*The statistical properties of entangled quantum states cannot be explained by any theory of local hidden variables.*

---

[10]Bell, J. S. 1964. *On the Einstein-Podolsky-Rosen paradox.* Physica 1:195-200

This result showed that Einstein's challenge to the existence of entanglement and his claim that quantum mechanics was an incomplete theory (the famous EPR paradox[11]) did not stand the test of time.

## 7.6  REFERENCES AND FURTHER READING

1. Schumacher B. and M. Westmoreland. 2010. *Quantum Processes, Systems, and Information.* Cambridge Univ. Press. Section 7.3.

2. Mermin, N. D. 2007. *Quantum Computer Science.* Cambridge Univ. Press. Sections 6.4 and 6.5.

3. Williams, C. P. 2011. *Explorations in Quantum Computing.* Springer-Verlag. Sections 11.5 and 11.6 and Chapter 12.

4. Kaye, Ph., R. Laflamme and M. Mosca. 2007. *An Introduction to Quantum Computing.* Oxford Univ. Press. Chapter 5.

## 7.7  EXERCISES AND PROBLEMS

1. Design two-qubit quantum circuits to create all four Bell states from classical product states.

2. When discussing teleportation, we used the Bell state $|B_{00}\rangle$ as the entangled resource shared by Alice and Bob. Find teleportation circuits appropriate for the other Bell states, namely, $|B_{01}\rangle$, $|B_{10}\rangle$, and $|B_{11}\rangle$.

3. Write the following two-qubit states in a Bell basis:

   (a) $|\Psi\rangle = \frac{1}{\sqrt{2}}(|00\rangle + i|11\rangle)$

   (b) $|\Psi\rangle = \frac{1}{\sqrt{2}}(|01\rangle + |11\rangle)$.

4. In quantum teleportation, Alice wants to send a one-qubit state $|\psi\rangle$ to Bob after they managed to share a Bell state. Prove the following relation, which is fundamental for a successful operation:

$$|\psi\rangle|\beta_{00}\rangle = \frac{1}{2}[|\beta_{00}\rangle|\psi\rangle + |\beta_{01}\rangle(\hat{X}|\psi\rangle) + |\beta_{10}\rangle(\hat{Z}|\psi\rangle) + |\beta_{11}\rangle(\hat{X}\hat{Z}|\psi\rangle)],$$

where $|\beta_{00}\rangle$, $|\beta_{01}\rangle$, $|\beta_{10}\rangle$, and $|\beta_{11}\rangle$ are the Bell states.

---

[11]Einstein, A., B. Podolsky, and N. Rosen. 1935. *Can quantum-mechanical description of physical reality be considered complete?.* Phys. Rev. 47:777-780

5. Using Qiskit, build a circuit mimicking the teleportation protocol. Prepare different single-qubit states $|\psi\rangle$ and show that the circuit produces the expected result.

6. Using Qiskit, build a circuit to implement superdense coding. Test all classical states Alice can convey to Bob.

7. Derive the expectation values shown in Eq. (7.5).

# CHAPTER 8

# Quantum Fourier Transform and Applications

Because of quantum interference, quantum computers are very efficient at finding and enhancing periodic structures in data. In 1994, Peter Shor introduced the first quantum algorithm that explored this property to solve a mathematical problem of great practical importance: factoring semi-prime numbers. To understand the algorithm (warning: it is not simple!), we need first to review some math.

## 8.1 FOURIER SERIES

Any periodic function $f$ on the real axis with period $L$ that is integrable over this period can be represented as a series of sines and cosines. If

$$f(x + L) = f(x)$$

for any $x \in \mathbb{R}$ and[1]

$$\left| \int_{-L/2}^{L/2} f(x) \, dx \right| < \infty,$$

then

$$f(x) = \sum_{n=0}^{\infty} a_n \cos\left(\frac{2\pi n x}{L}\right) + \sum_{n=1}^{\infty} b_n \sin\left(\frac{2\pi n x}{L}\right),$$

---

[1]Our condition for the convergence of the Fourier series is not a rigorous one. A better criterion exists and can be found in textbooks on harmonic analysis.

where the coefficients are given by

$$a_0 = \frac{1}{L} \int_{-L/2}^{L/2} f(x) \, dx,$$

$$a_n = \frac{2}{L} \int_{-L/2}^{L/2} f(x) \cos\left(\frac{2\pi n x}{L}\right) dx,$$

$$b_n = \frac{2}{L} \int_{-L/2}^{L/2} f(x) \sin\left(\frac{2\pi n x}{L}\right) dx,$$

with $n \geq 1$. It is common to name the coefficients $a_n$ the even ones, while the $b_n$ are called the odd ones. That is because the $a_n$ appear multiplying even functions (the cosines), while the $b_n$ multiply odd functions (the sines).

Let us check how this works for some periodic functions.

Take $f(x) = \cos(x)$. The period for this function is $L = 2\pi$. What are the Fourier coefficients in this case? The answer is simple (no need to perform integrations): by visual inspection, we can set $a_0 = 0$, $a_1 = 1$, $a_n = 0$ for $n \geq 2$, and $b_n = 0$ for $n \geq 1$.

Take now $f(x) = \sin^2(x)$. The period of this function is $L = \pi$. What are its Fourier coefficients? To find them, one can either carry out the integrations above or notice the following: $\sin^2(x) = [1 - \cos(2x)]/2$. Using the latter approach, we immediately identify $a_0 = \frac{1}{2}$, $a_1 = -\frac{1}{2}$, $a_n = 0$ for $n \geq 2$, and $b_n = 0$ for $n \geq 1$. Notice that the latter is expected since $\cos(x)$ is an even function and therefore cannot have odd functions in its decomposition.

Now consider the sawtooth function with period $2\pi$ and amplitude $\pi$ of Fig. 8.1, which is defined as

$$f(x + 2\pi k) = \frac{1}{\pi} \times \begin{cases} x, & -\pi < x < \pi \\ 0, & x = \pm\pi \end{cases} \quad \text{and} \quad k = 0, \pm 1, \pm 2, \ldots$$

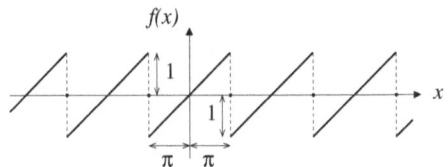

Figure 8.1  A sawtooth function.

Since it is an odd function, we can immediately set $a_n = 0$ for $n \geq 0$. Performing the integration over sines, we obtain $b_n = 2 \times (-1)^{n+1}/(\pi n)$ for $n \geq 1$. As result, we can write

$$f(x) = \frac{1}{\pi} \sum_{n=1}^{\infty} \frac{(-1)^{n+1}}{n} \sin(nx).$$

(Those who are familiar with electronic pulses, may recognize this decomposition.)

Often, we rearrange the Fourier series in a more compact form using exponential functions, namely,

$$f(x) = \sum_{n=-\infty}^{+\infty} c_n e^{-2\pi i n x/L}, \tag{8.1}$$

where the (possibly complex) coefficients are given by

$$c_n = \frac{1}{L} \int_{-L/2}^{L/2} f(x) e^{2\pi i n x/L} dx.$$

We will use this form from now on.

## 8.2   FOURIER TRANSFORM

Functions that do not have a finite periodic can also be expanded in terms of sines and cosines or exponentials, although for them one must substitute the infinite summation by an integral:

$$L \to \infty \quad \Longrightarrow \quad f(x) = \int_{-\infty}^{+\infty} \tilde{f}(k) e^{-2\pi i k x} dk,$$

where

$$\tilde{f}(k) = \int_{-\infty}^{+\infty} f(x) e^{2\pi i k x} dx.$$

The last equation is called the Fourier transform of $f(x)$.[2] It is very general and is applicable to periodic functions as well. For them, $\tilde{f}(k)$

---

[2]One often sees factors of $1/2\pi$ in one or the other expression. Usually, if the $2\pi$ factors are included in the exponential, there is no pressing need to also have the $1/2\pi$ scaling factors. However, these factors can also be absorbed in $\tilde{f}(k)$, provided that they are used consistently. For the sake of simplicity, we skip them altogether here.

will have maximum amplitude at values of $k$ corresponding to the inverse of the period of $f(x)$. Interestingly, if a function $f(x)$ is composed of various oscillating components, its Fourier transform will indicate the value of the inverse of all periods involved. This is often one of the most common uses of Fourier transforms.

## 8.3 DISCRETE FOURIER TRANSFORM

Let us suppose that a continuous function $f(x)$ is to be stored in a digital format. Obviously, in the processes of storing $f(x)$, we are not allowed to sample it continuously. Instead, we do it discretely, in equal increments $\Delta$ (the so-called sampling rate), such that

$$x \to x_q = q\Delta, \quad \text{with} \quad q = 0, 1, \ldots, N - 1.$$

Does this limit the range and granularity of the Fourier transform $\tilde{f}(k)$? The answer is yes, and the details are provided by the Nyquist-Shannon theorem, which states that, in this situation,

$$k \to k_n = \frac{n}{N\Delta}, \quad \text{with} \quad n = -\frac{N}{2}, \ldots, 0, \ldots, \frac{N}{2},$$

and

$$\tilde{f}(k_n) = \Delta \sum_{q=0}^{N-1} f_q e^{2\pi i n q / N}, \tag{8.2}$$

where

$$f_q = f(x_q).$$

The theorem basically says that the best resolution we can achieve in the $k$-space domain is $1/(N\Delta)$. Conversely, we cannot detect any structure or pattern (e.g., periodicity) in $f(x)$ that happens over ranges larger than $N\Delta$. Moreover, we cannot detect any oscillation faster than $\Delta$. We call the factor $1/(2\Delta)$ the Nyquist frequency or bandwidth.

The expression in Eq. (8.2) is an example of a discrete Fourier transform (DFT). As we will see later, they play an important role in cryptography, as well as in other areas of discrete data analysis. More generally, we define a DFT over a set of $N$ data points with complex amplitudes $\{f_q\}_{q=1,\ldots,N-1}$ as[3]

$$\tilde{f}_n = \frac{1}{\sqrt{N}} \sum_{q=0}^{N-1} f_q e^{2\pi i n q / N}, \tag{8.3}$$

where $n = 0, \ldots, N - 1$.

---

[3]The factor of $\sqrt{N}$ is a convenient way of making the expression of the inverse transform look the same way, except for a change in the sign of the exponent.

Notice that the DFT is nothing but a linear transformation between two sets of amplitudes. To see this more clearly, let us represent the amplitudes $\{f_q\}$ and $\{\tilde{f}_n\}$ as column vectors of length $N$ (i.e., matrices of dimensions $N \times 1$). Borrowing Dirac's notation, we write

$$|f\rangle = \frac{1}{\sqrt{N}} \begin{pmatrix} f_0 \\ f_1 \\ \vdots \\ f_{N-1} \end{pmatrix} \quad \text{and} \quad |\tilde{f}\rangle = \begin{pmatrix} \tilde{f}_0 \\ \tilde{f}_1 \\ \vdots \\ \tilde{f}_{N-1} \end{pmatrix},$$

leading to

$$|\tilde{f}\rangle = \hat{O}_{\text{DFT}}|f\rangle,$$

where the elements of the DFT $N \times N$ matrix $\hat{O}_{\text{DFT}}$ are defined as

$$[\hat{O}_{\text{DFT}}]_{nq} = e^{2\pi i n q/N}$$
$$= \lambda^{nq}.$$

Here, we introduced the phase factor

$$\lambda = e^{2\pi i/N},$$

which satisfies $\lambda^N = 1$. Notice that given a list of $N$ amplitudes $\{f_q\}$, $N^2$ multiplications are needed to compute the list of $N$ amplitudes $\{\tilde{f}_n\}$. We say that the computation complexity of computing $\tilde{f}$ is of order $O(N^2)$.

When we know that $f_q$ is periodic with period $r$, we can actually compute a closed expression for the Fourier transform amplitudes:

$$\tilde{f}_n = \frac{1}{\sqrt{N}} \sum_{q=0}^{N} f_q e^{2\pi i n q/N}$$

$$= \frac{1}{\sqrt{N}} \left[ \sum_{q=0}^{r-1} f_q e^{2\pi i n q/N} + \sum_{q=r}^{2r-1} f_q e^{2\pi i n q/N} + \ldots + \sum_{q=N-r}^{N-1} f_q e^{2\pi i n q/N} \right]$$

$$= \frac{1}{\sqrt{N}} \sum_{q=0}^{r-1} [f_q e^{2\pi i n q/N} + f_{q+r} e^{2\pi i n(q+r)/N} + \ldots + f_{q+N-r} e^{2\pi i n(q+N-r)/N}]$$

$$= \frac{1}{\sqrt{N}} \sum_{q=0}^{r-1} f_q e^{2\pi i n q/N} [1 + z + \ldots + z^{m-1}],$$

where $z = e^{2\pi i n r/N}$ and $m = N/r$. Notice the geometric series inside the square brackets:

$$1 + z + \ldots + z^{m-1} = \begin{cases} m & \text{if } z = 1 \\ (z^m - 1)/(z - 1) & \text{if } z \neq 1 \end{cases}.$$

Since $z^m = e^{2\pi i n} = 1$ for all $n$, we can put all these steps together and write

$$\tilde{f}_n = \begin{cases} 0, & \text{if } nr/N \text{ is not an integer} \\ \frac{m}{\sqrt{N}} \sum_{q=0}^{r-1} f_q\, e^{2\pi i n q/N}, & \text{if } nr/N \text{ is an integer} \end{cases} \qquad (8.4)$$

We managed to reduce the computation of the nonzero coefficients to a summation involving only data points within the period $r$. How does this summation behave? We can answer this question through an example.

*Example*: consider the following binary map $f : \{0,1\}^4 \to \{0,1\}$ where $f_0 = f_3 = 1$, $f_1 = f_2 = 0$, and $f_{q+r} = f_q$, with $r = 4$. This function is defined over $N = 2^4 = 16$ points and repeats the pattern $(1,0,0,1)$, going through a total of 4 cycles. See Fig 8.2.

Figure 8.2  Periodic function with period $r = 4$.

Employing Eq. (8.4), we find that $\tilde{f}_n$ is trivially zero for all values of $n$ but $0, 4, 8, 12$. For these four values, we obtain

$$\begin{aligned} \tilde{f}_0 &= 2 \\ \tilde{f}_4 &= 1 + e^{3\pi i/2} = \sqrt{2}\,e^{-\pi i/4} \\ \tilde{f}_8 &= 1 + e^{3\pi i} = 0 \\ \tilde{f}_{12} &= 1 + e^{\pi i/2} = \sqrt{2}\,e^{\pi i/4}. \end{aligned}$$

A convenient way to represent the DFT is via a "power spectrum", which is basically a graph of $|\tilde{f}_n|^2$ versus $n$, see Fig. 8.3.[4]

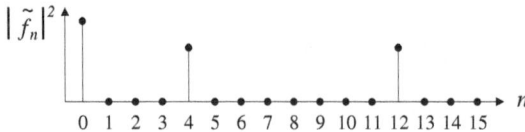

Figure 8.3  Power spectrum of the periodic function $f_q$ from Fig. 8.2.

The peaks in the power spectrum reveal information about the periodicity of the oscillations in $f_q$. The peak at $n = 0$ indicates that the

---

[4]Recall that the coefficients $\tilde{f}_n$ can be complex, which is indeed the case in this example.

function has a nonzero average value. The fact that there is a peak at $n = 4$ should come as no surprise since $N/r = 16/4 = 4$. Peaks at $n = 8$ and $n = 12$ are also expected because these values are integer multiples of 4. The relative amplitudes of these peaks are uniquely related to the structure of $f_q$ within a single period. Curiously, interference eliminates the $\tilde{f}_8$ coefficient (this is also a manifestation of the particular form of $f_q$ within a period).

From the DFT, we can extract a lot of information about $f_q$. Isolated, periodic peaks appear are the hallmark of a periodic $f_q$. If $f_q$ contains more than one periodic oscillation, then multiple nonperiodic peaks appear. When $f_q$ is not periodic, than nonzero amplitudes may appear for all values of $n$.

The best known classical algorithm for the computation of a DFT – the Fast Fourier Transform (FFT) – explores the structure of the matrix $\hat{O}_{DFT}$ to obtain the exact result with just $O(N \log N)$ steps when $N$ is large.

This seems pretty good, but when $N$ is itself exponentially large, as in $N = 2^n$, where $n$ is the number of bits needed to express $N$, it would take an enormous number of steps to implement an FFT. Unfortunately, such a situation is not uncommon, with $n$ ranging in the hundreds.

But a quantum computer can crack this problem with many fewer steps! Why? It is all about phases.

## 8.4   A QUANTUM ALGORITHM FOR FOURIER TRANSFORMS

To build a quantum algorithm to compute the DFT, let us begin by recalling the effect of a Hadamard gate on a generic one-qubit state.[5] Defining

$$|x\rangle = \alpha|0\rangle + \beta|1\rangle,$$

we have

$$\hat{H}|x\rangle = \frac{\alpha}{\sqrt{2}}(|0\rangle + |1\rangle) + \frac{\beta}{\sqrt{2}}(|0\rangle - |1\rangle).$$

Let us now assume that $x$ represents a classical bit, i.e., $x = 0$ ($\alpha = 1$ and $\beta = 0$) or $x = 1$ ($\alpha = 0$ and $\beta = 1$), in which case $|x\rangle$ is a classical state and we have

$$\hat{H}|x\rangle = \frac{1}{\sqrt{2}}|0\rangle + \frac{(-1)^x}{\sqrt{2}}|1\rangle.$$

---

[5]Coppersmith, D. 1994. *An Approximate Fourier Transform Useful in Quantum Factoring.* IBM Research Report RC 19642; arXiv:quant-ph/0201067

We can rewrite this expression in the following suggestive form:

$$\hat{H}|x\rangle = \frac{1}{\sqrt{2}} \sum_{y\in\{0,1\}} (-1)^{x\cdot y}|y\rangle, \tag{8.5}$$

with $x \in \{0,1\}$. The dot product in the exponent represents the bit-wise multiplication. This seems like an overkill of a notation, but hang in there. Notice that the Hadamard gate encodes information about $x$ in the relative phase factor between the two basis states.[6] It is straightforward to see that applying a second Hadamard gate decodes that information. A Hadamard gate essentially implements a one-bit DFT.

We can generalize this result for a register with $n$ qubits by applying a Hadamard gate to each individual qubit,

$$\hat{H}^{\otimes n}|x\rangle = \frac{1}{\sqrt{2^n}} \sum_{y\in\{0,1\}^n} (-1)^{x\cdot y}|y\rangle, \tag{8.6}$$

where $x \in \{0,1\}^n$. The dot product now describes a modulo-2 bit-wise scalar product: for $x = (x_0, x_1, \ldots, x_{n-1})$ and $y = (y_0, y_1, \ldots, y_{n-1})$, $x \cdot y = x_0 y_0 \oplus x_1 y_1 \oplus \cdots \oplus x_{N-1} y_{n-1}$.

The phase factor $-1 = e^{i\pi}$ appearing in Eqs. (8.5) and (8.6) is a special case of a more general one, $e^{2\pi i\omega}$.[7] In those equations, $\omega = 2^{-1}$ but imagine a state where $\omega$ could be any number in the range $[0 : 1)$. Such a state would then encode more information, similarly to $\tilde{f}_n$ in the DFT of Eq. (8.3).

Or, consider the following (inverted) problem: given an $n$-qubit quantum state with relative phases derived from a root phase factor $e^{2\pi i\omega}$,

$$|\Psi_n\rangle = \frac{1}{\sqrt{2^n}} \sum_{y=0}^{2^n-1} (e^{2\pi i\omega})^y|y\rangle = \frac{1}{\sqrt{2^n}} \sum_{y=0}^{2^n-1} e^{2\pi iy\omega}|y\rangle, \tag{8.7}$$

find $\omega$ (i.e., decode $\omega$ from the state) assuming $\omega \in [0;1)$. Such a problem is called a phase estimation.[8] It turns out that it is easier to solve this problem than the DFT one (at least for beginners), so let us present a

---

[6] This phase factor is smartly explored in Deutsch's and Deutsch-Jozsa's algorithms.

[7] Here, the real variable $\omega$ has nothing to do with the complex variable $\omega = e^{2\pi i/N}$ used in the previous section. We adopt the same Greek letter to stay close to the notation used in the literature.

[8] Kitaev, A. Yu. 1995. *Quantum measurements and the Abelian stabilizer problem.* arXiv:quant-ph/9511026

quantum algorithm for it and figure out a way to obtain the DFT based on the phase estimation solution.

Because $\omega < 1$, we can represent it in the binary form

$$
\begin{aligned}
\omega \quad &- \quad 0.\,\omega_1\,\omega_2\,\omega_3\cdots \\
&= \quad \omega_1\,2^{-1} + \omega_2\,2^{-2} + \omega_3\,2^{-3} + \cdots
\end{aligned}
$$

where $\omega_1 = 0, 1$, $\omega_2 = 0, 1$, $\omega_3 = 0, 1$, etc. This is very much analogous to the binary decomposition of integers, except that we are decomposing $\omega$ in powers of $2^{-1}$ and the exact decomposition may require an infinite number of terms. If you multiply $\omega$ by a positive power of 2, for instance $2^k$, you move the fraction point "." to the left, namely,

$$
2^k\omega = \omega_1\,\omega_2\,\omega_3\cdots\omega_k.\,\omega_{k+1}\,\omega_{k+2}\cdots
$$

Since $e^{2\pi i l} = 1$ for any integer $l$, we can write

$$
\begin{aligned}
e^{2\pi i(2^k\omega)} \quad &= \quad e^{2\pi i(\omega_1\,\omega_2\,\omega_3\cdots\omega_k.\,\omega_{k+1}\,\omega_{k+2}\cdots)} \\
&= \quad e^{2\pi i(\omega_1\,\omega_2\,\omega_3\cdots\omega_k)}e^{2\pi i(0.\omega_{k+1}\,\omega_{k+2}\cdots)} \\
&= \quad e^{2\pi i(0.\,\omega_{k+1}\,\omega_{k+2}\cdots)}.
\end{aligned}
\tag{8.8}
$$

Only the fractional part survives.

Thus, let us go back to the state in Eq. (8.7) and consider the case $n = 1$ with a single bit to represent the fractional part:

$$
\begin{aligned}
|\Psi_1\rangle \quad &= \quad \frac{1}{\sqrt{2}}\sum_{y=0}^{1} e^{2\pi i y\cdot(0.\omega_1)}|y\rangle \\
&= \quad \frac{1}{\sqrt{2}}\sum_{y=0}^{1} e^{2\pi i y\cdot\omega_1 2^{-1}}|y\rangle \\
&= \quad \frac{1}{\sqrt{2}}\sum_{y=0}^{1} e^{\pi i y\cdot\omega_1}|y\rangle \\
&= \quad \frac{1}{\sqrt{2}}\sum_{y=0}^{1} (-1)^{y\cdot\omega_1}|y\rangle,
\end{aligned}
$$

which is the form expected when a Hadamard gate acts on a qubit in the classical state $|\omega_1\rangle$. We can then apply a second Hadamard to retrieve the value of $\omega_1$ (since $\hat{H}^{-1} = \hat{H}$):

$$
\begin{aligned}
\hat{H}|\Psi_1\rangle &= \hat{H}\frac{1}{\sqrt{2}}\sum_{y=0}^{1}(-1)^{y\cdot\omega_1}|y\rangle \\
&= \hat{H}\frac{1}{\sqrt{2}}[|0\rangle + (-1)^{\omega_1}|1\rangle] \\
&= |\omega_1\rangle.
\end{aligned}
$$

By measuring the qubit after the Hadamard gate has been applied, we can obtain $\omega_1$ and then estimate $\omega$ up to one bit of precision.

If we want more precision, we simply add more bits. Suppose $\omega = 0.\omega_1\omega_2$ (two bits of precision). Then,

$$
\begin{aligned}
|\Psi_2\rangle &= \frac{1}{2}\sum_{y=0}^{3}e^{2\pi iy\cdot(0.\omega_1\omega_2)}|y\rangle \\
&= \frac{1}{2}\sum_{y_1=0}^{1}\sum_{y_2=0}^{1}e^{2\pi i(y_2\,2^1+y_1\,2^0)\cdot(0.\omega_1\omega_2)}|y_1\rangle|y_2\rangle \\
&= \frac{1}{2}\sum_{y_1=0}^{1}\sum_{y_2=0}^{1}e^{2\pi iy_2(0.x_2)}e^{2\pi iy_1(0.\omega_1\omega_2)}|y_1\rangle|y_2\rangle \\
&= \frac{1}{2}\left(\sum_{y_1=0}^{1}e^{2\pi iy_1(0.\omega_1\omega_2)}|y_1\rangle\right)\left(\sum_{y_2=0}^{1}e^{2\pi iy_2(0.\omega_2)}|y_2\rangle\right) \\
&= \frac{1}{\sqrt{2}}(|0\rangle + e^{2\pi i(0.\omega_1\omega_2)}|1\rangle)\frac{1}{\sqrt{2}}(|0\rangle + e^{2\pi i(0.\omega_2)}|1\rangle) \\
&= \frac{1}{\sqrt{2}}(|0\rangle + e^{2\pi i(0.\omega_1\omega_2)}|1\rangle)\frac{1}{\sqrt{2}}[|0\rangle + (-1)^{\omega_2}|1\rangle].
\end{aligned}
$$

We can thus extract $\omega_2$ by applying a Hadamard on the second qubit and measuring it. Since $|\Psi_2\rangle$ is a product state of two one-qubit states,

$$
|\Psi_2\rangle = |\psi_{(1)}\rangle \otimes |\psi_{(2)}\rangle,
$$

such an operation will not affect the state of the first qubit. If we determine that $\omega_2 = 0$, then we apply a Hadamard gate on the first qubit, measure it, and retrieve $x_1$ directly. If we find that $\omega_2 = 1$ instead, a Hadamard followed by a measurement of the first qubit will not be sufficient to determine $\omega_1$. In this case, we need to first apply a phase gate $\hat{U}_2$ with a proper phase,

$$
\hat{U}_2 = \begin{pmatrix} 1 & 0 \\ 0 & e^{-2\pi i2^{-2}} \end{pmatrix} = \begin{pmatrix} 1 & 0 \\ 0 & e^{-2\pi i(0.01)} \end{pmatrix},
$$

to the first qubit:

$$\begin{aligned}
|\psi'_{(1)}\rangle &= \hat{U}_2|\psi_{(1)}\rangle \\
&= \hat{U}_2 \frac{1}{\sqrt{2}}(|0\rangle + e^{2\pi i(0.\,\omega_1 1)}|1\rangle) \\
&= \frac{1}{\sqrt{2}}(|0\rangle + e^{2\pi i(0.\,\omega_1 1)}e^{-2\pi i(0.01)}|1\rangle) \\
&= \frac{1}{\sqrt{2}}(|0\rangle + e^{2\pi i(0.\,\omega_1)}|1\rangle) \\
&= \frac{1}{\sqrt{2}}[|0\rangle + (-1)^{\omega_1}|1\rangle].
\end{aligned}$$

Now a Hadamard gate followed by a measurement on the first qubit will do the trick and reveal $\omega_1$. The diagram in Fig. 8.4 is the complete circuit that implements these operations (notice that we are displaying bits from bottom up).

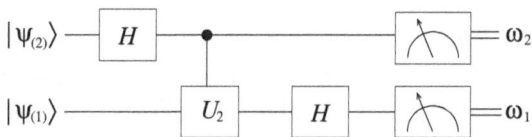

Figure 8.4  Phase estimation circuit with two-bit precision.

Can we generalize this approach where we want three bits of precision on the estimation of $\omega$? Certainly. Check the circuit in Fig 8.5, where with $\omega = 0.\,\omega_1\,\omega_2\,\omega_3$ and

$$\begin{aligned}
|\psi_{(1)}\rangle &= \frac{1}{\sqrt{2}}(|0\rangle + e^{2\pi i(0.\,\omega_1\,\omega_2\,\omega_3)}|1\rangle) \\
|\psi_{(2)}\rangle &= \frac{1}{\sqrt{2}}(|0\rangle + e^{2\pi i(0.\,\omega_2\,\omega_3)}|1\rangle) \\
|\psi_{(3)}\rangle &= \frac{1}{\sqrt{2}}(|0\rangle + e^{2\pi i(0.\,\omega_3)}|1\rangle).
\end{aligned}$$

In this case, we made use of the following single-qubit phase gates:[9]

$$\hat{U}_k = \begin{pmatrix} 1 & 0 \\ 0 & e^{-2\pi i 2^{-k}} \end{pmatrix},$$

with $k = 2, 3$.

---

[9]It is straightforward to show that a phase gate is equivalent to an $R_z$ rotation gate up to an overall phase factor.

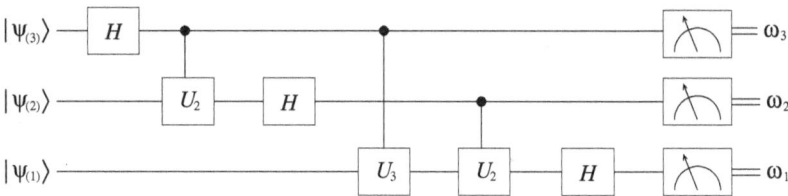

Figure 8.5 Phase estimation circuit with three-bit precision.

The fact that we are displaying bits bottom up may not be pleasant to the eye. In fact, before the layer of measurements, the three-qubit system is in the state $|\omega_3 \omega_2 \omega_1\rangle$, rather than the standard $|\omega_1 \omega_2 \omega_3\rangle$. There is a simple way to circumvent this shortcoming: apply a series of SWAP gates.[10]

When we want to go further in the precision of the phase estimation and extend it to $n$ bits, say $\omega = 0.\omega_1 \omega_2 \cdots \omega_{n-1}$, it is straightforward to generalize the circuit used for three bits.

In general, the number of gates needed to implement this algorithm is $O(n^2)$, which is a very reasonable computation complexity cost.

This algorithm allowed us to solve the following problem: given

$$|\Psi_n\rangle = \frac{1}{\sqrt{2^n}} \sum_{y=0}^{2^n-1} e^{2\pi i \omega \cdot y} |y\rangle, \tag{8.9}$$

find $\omega$ up to $n$ bits of precision using $O(n^2)$ computational steps and a reversible quantum circuit. Good!

But let us consider the inverse of this problem, namely, given an $n$-qubit classical state $|\omega\rangle$, find its DFT, which is basically what Eq. (8.9) represents [recall Eq. (8.3) and set $N = 2^n$]. Therefore, we can conclude that it is possible to compute the DFT of a set of $N$ data points by applying the *inverse* of the phase estimation circuit! Because the circuit is essentially a unitary operation, it is always reversible. And we can do so by incurring only in a computational cost of $O(n^2)$, with $n = \log_2 N$. That is a huge (exponential!) speed up in comparison to FFT and any other known classical algorithm! We call the inverse of the phase estimation algorithm a quantum Fourier transform (QFT). QFT has computational complexity $O(n^2)$ whereas FFT has $O(n\,2^n)$, thus QFT offers an exponential speedup.

---

[10]While this seems innocuous at first, it does have some implications for the classical simulability of such quantum circuits! SWAP gates can increase entanglement, which is always difficult to handle in a classical computer.

The diagram in Fig. 8.6 shows a circuit that implements the QFT algorithm.

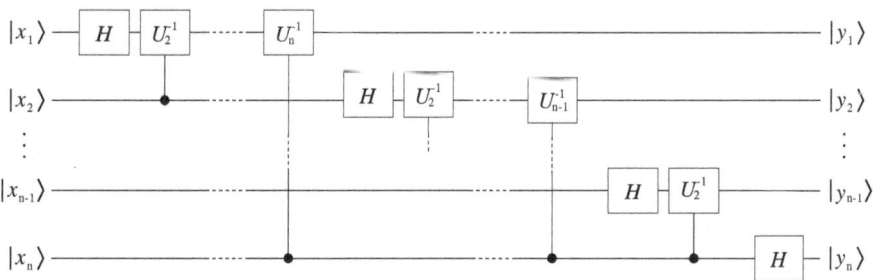

Figure 8.6 QFT circuit.

There are a few important points to make:

1. Notice that the phase gates appear inverted in comparison to the phase estimation circuit. The Hadamard operators are also inverted, but since $\hat{H}^{-1} = \hat{H}$ they show up the same.

2. This circuit can only compute the DFT (or QFT) when $N = 2^n$, i.e., when $x$ is exactly represented by $n$ bits. This is because the phase estimation circuit retrieves phases that are an integer multiple of $1/2^n$.

3. As we will discuss shortly, we can retrieve phases using this algorithm even when they are not exact multiples of $1/2^n$. We can do so with accuracy $O(1/2^n)$ when we use $n$ qubits.

4. Since our QFT algorithm relies entirely on unitary gates, the QFT itself can be thought of as an $n$-qubit gate:

$$\hat{U}_{\text{QFT}}|\omega\rangle_n = \frac{1}{\sqrt{2^n}} \sum_{y=0}^{2^n-1} e^{2\pi i \omega \cdot y}|y\rangle_n. \qquad (8.10)$$

Let us return to the case when the phase we want to determine is not an exact multiple of $1/2^n$:

$$|\psi\rangle = \frac{1}{\sqrt{2^n}} \sum_{y=0}^{2^n-1} e^{2\pi i \omega \cdot y}|y\rangle,$$

where $\omega \in [0; 1]$. How do we find $\omega$?

Upon applying the phase estimation algorithm to $|\psi\rangle$ via a QFT, one obtains the state

$$\hat{U}_{\text{QFT}}^{-1}|\psi\rangle = \sum_x \alpha_x(\omega)|x\rangle,$$

where $|\alpha_x(\omega)|^2$ is the probability of the outcome $x$. Calling $\tilde{\omega} = x/2^n$, we can say that $\tilde{\omega}$ is a good approximation to $\omega$ with probability $|\alpha_x(\omega)|^2$ peaked around the integer $x$ such that $\tilde{\omega}$ is the closest multiple integer of $1/2^n$ to $\omega$. One can prove that this is true with probability at least $4/\pi^2$ (about 40.5%). This result guarantees that we can determine $\omega$ with accuracy $O(1/2^n)$ when we use $n$ qubits:

$$|\tilde{\omega} - \omega| \le \frac{1}{2^{n+1}}, \quad \text{prob} = \frac{4}{\pi^2}.$$

## 8.5 APPLICATION OF QFT: FINDING PERIODS AND CRACKING RSA

Let us define a periodic state as

$$|\phi_r; b\rangle = \frac{1}{\sqrt{m}} \sum_{j=0}^{m-1} |jr + b\rangle,$$

where $r$ is the period, $b$ is a shift, and $m$ is the number of repetitions. Assume that all these parameters are integers and $0 \le b \le r - 1$. Given $L = m \times r$ (the range of the state), can we find $r$?

One may be tempted to use a QFT to extract $r$, similarly to how we would do for a continuous function (we expect that a QFT transform will have a peak at $x = r$). However, the problem is more subtle.

Let

$$\hat{U}_L^{\text{QFT}}|x\rangle = \frac{1}{\sqrt{L}} \sum_{y=0}^{L-1} e^{-2\pi i x \cdot y/L}|y\rangle,$$

where $\{|0\rangle, |1\rangle, \ldots, |L-1\rangle\}$ are basis states. Then,

$$
\begin{aligned}
\hat{U}_{mr}^{\text{QFT}}|\phi_r; b\rangle &= \frac{1}{\sqrt{m}} \sum_{j=0}^{m-1} \hat{U}_{mr}^{\text{QFT}}|jr + b\rangle \\
&= \frac{1}{\sqrt{m}} \sum_{j=0}^{m-1} \frac{1}{\sqrt{mr}} \sum_{y=0}^{mr-1} e^{-2\pi i (jr+b)y/mr}|y\rangle \\
&= \frac{1}{m\sqrt{r}} \sum_{y=0}^{mr-1} e^{-2\pi i\, by/mr} \left( \sum_{j=0}^{m-1} e^{-2\pi i\, jy/m} \right) |y\rangle.
\end{aligned}
$$

Notice that

$$\sum_{j=0}^{m-1} e^{-2\pi i\, jy/m} = \begin{cases} m, & \text{if } y \text{ is a multiple of } m \\ 0, & \text{otherwise destructive interference} \end{cases}.$$

Therefore,

$$\hat{U}_{mr}^{\mathrm{QFT}}|\phi_r;b\rangle = \frac{1}{\sqrt{r}} \sum_{k=0}^{r-1} e^{-2\pi ibk/r}|km\rangle$$

(we set $y = km$). If we measure the resulting state in the computational basis we will obtain $km$ for some value of $k = 0, 1, \ldots, r-1$. Since we know $mr$, we could in principle compute

$$\frac{km}{mr} = \frac{k}{r}$$

to retrieve $r$. However, this only works when $k$ and $r$ have no common factor. For instance, let $mr = 60$ and $km = 24$. Then, $km/mr = k/r = 24/60 = 12/30 = 6/15 = 2/5$. Is $r = 60$, 30, 15, or 5? You cannot tell!

One way to pin down $r$ is to repeat the procedure many times, so that at least once we hit a value of $k$ that does not have a common factor with $r$. One can show that the number of repetitions needed is $O(\log \log r)$, which is not much. However, the real cost of this method is the number of steps in the execution of the QFT, which is $O(n^2)$, which, in this context, is equivalent to $O(\log^2 r)$.

Good! What about the situation when we do not know $m$? Namely, when

- $n$ is known (or given) and

- a black box generates the state

$$|\phi_r;b\rangle = \frac{1}{\sqrt{M}} \sum_j |jr + b\rangle,$$

where $0 < jr + b \leq 2^n$ and $M$ is some constant value that makes the state normalized.

How do we find $r$ in this case? Answer: by using the inverse of a QFT, i.e., phase estimation! We can obtain a value $x$ such that $x/2^n$ is close to $k/r$ for some integer $k \in \{0, 1, 2, \ldots, r\}$. How close? We will state the answer without proving it: it is

$$\left|\frac{x}{2^n} - \frac{k}{r}\right| \leq \frac{1}{2Mr}.$$

The probability of obtaining such a value of $x$ is at least $(M/2^n)(4/\pi^2)$.

Using this result and a technique called continuous fraction, we can arrive at $r$ with high probability and accuracy. We thus reduced the computation cost of the solution from $O(2^n)$ steps to $O(n^2)$ steps. That is an exponential speed up!

The period finding algorithm just described was invented by Peter Shor in 1994 to solve another problem: factoring of large semi-prime numbers.[11] This is an extremely important application because the most common asymmetric cryptosystem in use today is based on the presumed difficulty to factor semi-prime numbers, the so-called RSA encryption protocol (RSA stands for the name of the inventors).[12] To understand this application, we need to delve in cryptography.

### 8.5.1 RSA and period finding

RSA is an example of a public-private key cryptosystem (also known as asymmetric encryption). Imagine that Alice wants people to send her encrypted messages that only she can decrypt. Here is a procedure for realizing this tasks:

- Alice takes two large odd primes, $p$ and $q$, and computes $n = pq$.

- Alice also chooses an integer $e$ such that $1 < e < (p-1)(q-1)$.[13]

- Alice computes $d$ such that $de=1 \bmod (p-1)(q-1)$.[14, 15]

- Alice broadcasts her public key $(n, e)$ and keeps $d$ private.

If Bob wants to send a message to Alice, he uses her public key to encode it. Calling Bob's plaintext message $N$ and its ciphertext (encrypted) version $C$, Bob can compute $C$ by exponentiating $N$ $e$ times, modulo $n$. Namely,

$$C = N^e \bmod n.$$

(Modular exponentiation can be done quite efficiently in a classical computer.)

---

[11]Shor, P. W. 1994. *Algorithms for quantum computation: discrete logarithm and factoring.* Proceedings of the 35th Annual Symposium on Foundation of Computer Science, Santa Fe, NM, USA: 124-134

[12]Rivest, R., A. Shamir, L. Adleman. 1978. *A method for obtaining digital signatures and public-key cryptosystems.* Commun. ACM 21: 120-126

[13]$e$ should be a co-prime of $(p-1)(q-1)$.

[14]$d$ is the inverse of $e$, modulo $(p-q)(q-1)$.

[15]If you are not familiar with modular operations, consider the example $7 \times 2 \bmod 5 = 14 \bmod 5 = (2 \times 5 + 4) \bmod 5 = 1$.

Once Alice receives Bob's ciphertext $C$, she can retrieve the plaintext $N$ by performing the operation

$$N = C^d \bmod n.$$

Because only Alice has the private key $d$, she is the only person who can retrieve the plaintext $N$ efficiently. Anyone else would have to extract $p$ and $q$ from $n$, i.e., would need to factorize $n$ into its prime factors, and then compute $d$ from $e$, $p$, and $q$ (this last part is easy). Since $n$ typically has hundreds of decimal digits, it can reach thousands of bits in size. All known classical algorithms to factor $n$ require an exponential number of operations to perform factoring. The best one, the sieve algorithm, scales as $O\left(e^{\sqrt{\ln n \ln \ln n}}\right)$.

It turns out that factoring a semi-prime number can be reduced to finding the periodicity of a function, thanks to Miller's algorithm. Let $f(y) = x^y \bmod n$ for a given $x < y$, where $x$ and $n$ have no common factor. If you can find the smallest $r$ such that $f(r) = 1$, then the prime factors of $n$ are the greatest common divisors of $(x^{r/2}+1)$ and $(x^{r/2}-1)$ with respect to $n$. But solving for $f(r) = 1$ is the same as finding the period of $f(y)$: since

$$
\begin{aligned}
f(y + r) &= x^{y+r} \bmod n \\
&= [(x^y \bmod n)(x^r \bmod n)] \bmod n,
\end{aligned}
$$

if $x^r \bmod n = 1$ then

$$
\begin{aligned}
f(y + r) &= [x^y \bmod n] \bmod n \\
&= x^y \bmod n \\
&= f(y).
\end{aligned}
$$

Therefore, $r$ is the period of $f(y)$.

We have a powerful quantum algorithm to find $r$, one which requires only $O(n^2)$ steps!

Therefore, a quantum computer capable of implementing a QFT calculation over thousands of qubits (that is a lot of qubits!) would crack RSA. This discovery by Peter Shor set in motion a race to realize quantum computing in practice. It also attracted the attention of multiple funding agencies, especially defense ones, which were eager to explore and acquire such a technology.

The current generation of quantum computers is still very far from having the thousands of qubits needed to accommodate the number bits used in the RSA cryptosystem, but it is possible that in the near future these machines will reach that limit. In that case, RSA will be trivially broken and this is a very troubling scenario because adversaries and bad actors may be already collecting sensitive RSA-encrypted data today to decrypt them once QFT is readily available.[16] This has lead government agencies around the globe to pursue new encryption schemes that do not rely on RSA and are "post-quantum" secure. In the USA, NIST has recently selected a number of new cryptosystems that are believed (but not rigorously proven) to fall into this category.[17]

## 8.6   OTHER APPLICATIONS OF QFT

### 8.6.1   Eigenvalue estimation

Another application of QFT where there is an exponential speed up in relation to known classical algorithms is the estimation of eigenvalues of a unitary operator that has a known eigenvector. Let us define the problem more precisely.

Find $\omega$ in

$$\hat{U}|\Psi\rangle = e^{2\pi i\omega}|\Psi\rangle$$

for $0 < \omega < 1$ when both $\hat{U} = \hat{U}^\dagger$ and $|\Psi\rangle$ are known.

Notice that simply applying $\hat{U}$ to $|\Psi\rangle$ does not allow one to retrieve $\omega$ since it shows up in the exponent of a global phase factor of $|\Psi\rangle$. We need another strategy.

The basic idea is to create a superposition state where the phase factor $e^{2\pi i\omega}$ appears as a relative phase.

Let $\hat{U}$ be an $n$-qubit unitary gate and $\hat{C}_U$ its control version (the control is performed by an extra, ancillary qubit). Set the control qubit initially to 1 and apply $\hat{C}_U$:

$$\begin{aligned}
\hat{C}_U|1\rangle \otimes |\Psi\rangle &= |1\rangle \otimes \hat{U}|\Psi\rangle \\
&= |1\rangle \otimes e^{2\pi i\omega}|\Psi\rangle \\
&= e^{2\pi i\omega}|1\rangle \otimes |\Psi\rangle.
\end{aligned}$$

---

[16]In practice, one often uses RSA just to perform key exchanges, while actual data encryption is performed by other symmetric methods that are not known to be crackable, such as AES. But commercial transactions and even most cryptcurrencies are completely reliant on RSA or variants that are also crackable by a powerful-enough quantum computer.

[17]https://csrc.nist.gov/projects/post-quantum-cryptography

You can see that the phase factor pops up. Now, use instead a superposition state for the control qubit:

$$\hat{C}_U(\alpha|0\rangle + \beta|1\rangle) \otimes |\Psi\rangle = (\alpha|0\rangle + \beta e^{2\pi i \omega}|1\rangle) \otimes |\Psi\rangle.$$

The phase factor $e^{2\pi i \omega}$ is now a relative phase. It can be retrieved via a quantum interference procedure. What procedure? Phase estimation! (i.e., the inverse of QFT.) Let us go through the steps.

First you use QFT to build an appropriate superposition state. Then you use that state to control $\hat{U}$. Finally, you use the inverse of QFT to retrieve the phase. All these steps combined yield the circuit shown in Fig. 8.7:

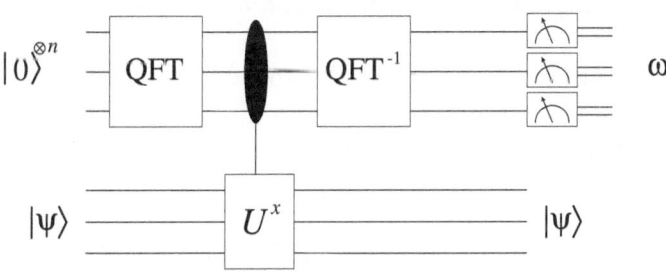

Figure 8.7 Eigenvalue estimation circuit.

Here, the multi-qubit operator $\hat{U}^x$ represents a sequence of $\hat{C}_U$ gates, each one being a descending power of the original $\hat{U}$ operator. For example: $\hat{U}^{2n-1}, \hat{U}^{2n-2}, \ldots, \hat{U}$.

### 8.6.2 Discrete logarithms

This is another problem relevant to cryptography in which quantum computers can have an edge. Let us first define the discrete logarithm problem.

Given $a$ and $b$ in $Z_p^*$ such that $a = b^t$ where $t$ is an integer from the set $\{0, 1, \ldots, r - 1\}$ and $r$ is of the order of $a$, find $t$.

$t$ is called the discrete logarithm of $b$. Here, $Z_p^*$ denotes the set of integers that relate to each other via addition and multiplication modulo $p$, such that any element of this set is co-prime with $p$. $Z_p^*$ combined with such a multiplication is called the multiplicative group of ring $Z_p$.

This is clearly a very abstract mathematical problem, but is at the core of some cryptosystems. Security of these systems is based on the difficulty of finding $t$ when $a$ and $b$ are large (i.e., when many bits are required to represent these two numbers). It turns out that the phase estimation algorithm can be used to solve this problem as well. The protocol is similar to that used in the eigenvalue estimation problem but more control registers are needed. We will not discuss it further in this book, but you should be aware of this application.

### 8.6.3   Hidden subgroup problem

There is a more general class of problems that can be tackled with super-polynomial speed up with a quantum computer capable of implementing a QFT: the hidden subgroup problem. Let us define it.

Let $f : G \to X$ where $G$ is a group and $X$ represents a finite set. Let $S \in G$ ($S$ is a subgroup of $G$) such that for any $x, y \in G$, $f(x) = f(y)$ if and only if $x + S = y + S$. Find $S$.

This is a generalization of Simon's problem discussed in Ch. 6.

It turns out that all previous problems we discussed (and a few more that we have not) can be cast as hidden subgroup problems! Thus, this is an essential class of problem as far as quantum algorithms are concerned, albeit its abstract definition.

The generic protocol that tackles the hidden subgroup problem is beyond the scope of this book![18]

## 8.7   REFERENCES AND FURTHER READING

1. James, J. F. 2011. *A Student's Guide to Fourier Transforms: with Applications in Physics and Engineering*, 3rd edition. Cambridge Uni. Press. Chapters 1 and 9.

2. Mermin, N. D. 2007. *Quantum Computer Science*. Cambridge Univ. Press. Chapter 3.

3. Nielsen M. A. and I. L. Chuang. 2000. *Quantum Computation and Quantum Information*. Cambridge Univ. Press. Chapter 5.

4. Kaye, Ph., R. Laflamme and M. Mosca. 2007. *An Introduction to Quantum Computing*. Oxford Univ. Press. Chapter 7.

5. Williams, C. P. 2011. *Explorations in Quantum Computing*. Springer-Verlag. Chapter 6.

---

[18]Ettinger, M., P. Hoyer, and E. Knill. 2004. *The quantum query complexity of the hidden subgroup problem is polyonimal*. Info. Process. Lett. 91: 43-48

## 8.8 EXERCISES AND PROBLEMS

1. Prove that these two representations of a Fourier series (using cosines and sines or exponentials) are equivalent. What property do the $c_n$ coefficients in Eq. (8.1) need to satisfy for this to be true?

2. Show that $|\Psi_n\rangle$ in Eq. (8.7) is a product state.

3. Starting from Eq. (8.10), find $\hat{U}_{\mathrm{QFT}}^{-1}$.

4. Show that $\hat{U}_{\mathrm{QFT}} = \hat{H}$ for $n = 1$.

5. Using Qiskit, build a quantum circuit that implements a QFT (Quantum Fourier Transform) for a register containing four qubits. Your construction should include the following steps:

   - qubits numbered from 0 to 3;
   - an ordered list of gates (including gate type and which qubits the gate acts on);
   - a diagram showing all the gates in the circuit;
   - a brief explanation of the role played by all stages of the circuit;
   - the Qiskit code defining the circuit;
   - results of a few experiments to verify that the output probability distribution is consistent with what is expected from a QFT.

6. Based on the results you obtained in problem 3, build a circuit that implements the inverse QFT for a register containing four qubits. Your construction should include the following steps:

   - qubits numbered from 0 to 3;
   - an ordered list of gates (including gate type and which qubits the gate acts on);
   - a diagram containing all the gates in the circuit;
   - the Qiskit code defining the circuit;
   - results of a a few experiments to verify that the output probability distribution is consistent with what is expected from an inverse QFT.

7. Show that the circuits obtained for problems 3 and 4 (QFT and its inverse), when concatenated, produce the equivalent of an identity circuit. *Hint*: run experiments and show that, for any input state, an output state is equal to the input state.

# CHAPTER 9

# Quantum Search and Applications

There is another class of quantum algorithms that do not rely on quantum phase estimation. For algorithms in this other class, the speedup is only polynomial with respect to classical algorithms. However, they tackle problems of great practical importance and are of wider relevance. Their paradigm is the solution of a search of an unstructured database, which is performed via quantum amplitude amplification.

## 9.1   GROVER'S ALGORITHM

Suppose you have a set of $N$ unsorted books and you want to find one or more books that have the word "We" on its first page. How many trials would it take you to find at least one book with the desired property? If you are very lucky, one trial (i.e., the very first book you pick); if you are very unlucky and there is only one book with such a property in the set, $N$ trials. On average, it should take about a fraction of $N$ trials (more precisely, $N/2$ trials if there is only one book with the desired property). Thus, it seems that any attempt to solve this problem will have to go through $O(N)$ trials in general.

It turns out that no known classical algorithm can do better than $O(N)$. However, Lov Grover showed in 1996 that a quantum computer can solve this class of problems with just $O\left(\sqrt{N}\right)$ trials.[1] That is a polynomial speed up in relation to classical algorithms.

---

[1]Grover, L. K. 1996. *A fast quantum mechanical algorithm for database search.* Proceedings of the 28th annual ACM symposium on theory of computing: 212-219

Grover's algorithm solves the following generic problem: let $\hat{U}_f$ be a black-box operator (i.e., an oracle, using the language of computer science) that computes $f(x)$ given $x$, where $f : \{0,1\}^n \to \{0,1\}$. Find $x$ such that $f(x) = 1$.

In the worst-case scenario, the algorithm requires $O\left(\sqrt{2^n}\right)$ queries to the oracle, namely, the application of the black-box operator. It explores quantum parallelism and amplitude amplification. But how does it work?

To understand it, for simplicity, let us assume that $f(x) = 1$ for only one (unknown) value $x = \omega$. We start with an $n$-qubit register for $x$ and a single target qubit $y$. Then, applying the black-box operator $\hat{U}_f$ we find

$$\hat{U}_f |x\rangle |y\rangle = |x\rangle |y \oplus f(x)\rangle$$

(this should be familiar to those who recall Deutsch's algorithm). If we set $y = 0$ initially, then

$$\hat{U}_f |x\rangle |0\rangle = |x\rangle |f(x)\rangle.$$

Applying the operator $\hat{U}_f$ on the initial state $|x\rangle |0\rangle$ is a way to query the oracle (through a measurement of the target qubit) but does not yet yield $\omega$. For that, let us prepare the control register in a superposition of all possible values. We can do that by applying a sequence of Hadamard gates, $\hat{H}^{\otimes n}$, one for each qubit in the control register, initially set to $|0\rangle^{\otimes n}$, yielding

$$\frac{1}{\sqrt{2^n}} \sum_{x=0}^{2^n-1} |x\rangle.$$

(We employ the superscript $n$ to remind us that this is an $n$-qubit state vector.) Formally, we can split this superposition into two parts: one that contains the solution and another that does not, namely,

$$\frac{1}{\sqrt{2^n}} |\omega\rangle + \sqrt{\frac{2^n-1}{2^n}} |\Omega\rangle,$$

where

$$|\Omega\rangle = \frac{1}{\sqrt{2^n-1}} \sum_{x \neq \omega} |x\rangle.$$

Now, let us apply the operator $\hat{U}_f$ to this superposition:

$$\hat{U}_f \left( \frac{1}{\sqrt{2^n}} |\omega\rangle + \sqrt{\frac{2^n-1}{2^n}} |\Omega\rangle \right) |0\rangle = \frac{1}{\sqrt{2^n}} |\omega\rangle |1\rangle + \sqrt{\frac{2^n-1}{2^n}} |\Omega\rangle |0\rangle.$$

If we now measure the target qubit we will get 1 with probability $1/2^n$, and the control register will contain the solution $\omega$. This is not yet very satisfying (since the probability of success is exponentially low) and does not differ much from the ordinary trial-and-error classical approach.

But we can do better by first putting the target qubit in a superposition. Why? Check it out:

$$
\begin{aligned}
\hat{U}_f|x\rangle(\hat{H}|1\rangle) &= \hat{U}_f|x\rangle\left(\frac{|0\rangle - |1\rangle}{\sqrt{2}}\right) \\
&= \frac{1}{\sqrt{2}}(\hat{U}_f|x\rangle|0\rangle - \hat{U}_f|x\rangle|1\rangle) \\
&= \frac{1}{\sqrt{2}}(|x\rangle|f(x)\rangle - |x\rangle|f(x)\oplus 1\rangle) \\
&= \frac{1}{\sqrt{2}}(-1)^{f(x)}(|x\rangle|0\rangle - |x\rangle|1\rangle) \\
&= (-1)^{f(x)}|x\rangle(\hat{H}|1\rangle).
\end{aligned}
\tag{9.1}
$$

Encoding the target qubit in a superposition state produces a phase shift upon acting with the black-box operator $\hat{U}_f$ on the $(n+1)$-qubit state vector. Looking at Eq. (9.1), it is clear that we can associate this phase shift to the control register only, namely,

$$
\hat{U}_f|x\rangle = (-1)^{f(x)}|x\rangle,
$$

so long as we use $\hat{H}|1\rangle$ to prepare the target qubit. From now on, we will implicitly assume that the target qubit is in that superposition state and will drop it from the state vector expressions. We will only explicitly indicate the vector associated to the control register.

It is convenient to introduce an auxiliary operator to take care of the phase shift. Consider an operator $\hat{P}_0$ such that

$$
\hat{P}_0|x\rangle = \begin{cases} -|x\rangle, & x \neq 0 \\ |0\rangle^{\otimes n}, & x = 0 \end{cases}.
$$

Let us now use this operator to build another one, which incorporates the sequence of Hadamard gates and the black-box operator,

$$
\hat{G} = \hat{H}^{\otimes n}\,\hat{P}_0\,\hat{H}^{\otimes n}\,\hat{U}_f
$$

(the first operator to act is the black-box one and the last is a Hadamard battery). A diagrammatic representation of $G$ is shown in Fig. 9.1.

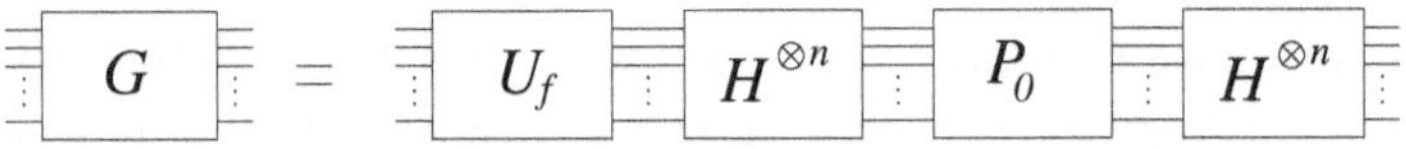

Figure 9.1  The Grover iterate.

This operator is called the Grover iterate and it amplifies the probability of finding $w$. To see that, consider the following steps.

First, define two state vectors,

$$|\psi\rangle = \hat{H}^{\otimes n}|0\rangle^{\otimes n}$$

and

$$|\phi\rangle = \hat{H}^{\otimes n}|x\rangle \quad \text{for } x \neq 0 \text{ only,}$$

and notice that these two vectors are orthogonal to each other,

$$\langle\phi|\psi\rangle = 0$$

for all $|\phi\rangle$ (there are $2^n - 1$ such $|\phi\rangle$ vectors).

Second, notice that

$$
\begin{aligned}
\hat{H}^{\otimes n}\hat{P}_0\,\hat{H}^{\otimes n}|\psi\rangle &= \hat{H}^{\otimes n}\,\hat{P}_0|0\rangle^{\otimes n} \\
&= \hat{H}^{\otimes n}|0\rangle^{\otimes n} \\
&= |\psi\rangle
\end{aligned}
$$

and

$$
\begin{aligned}
\hat{H}^{\otimes n}\,\hat{P}_0\,\hat{H}^{\otimes n}|\phi\rangle &= \hat{H}^{\otimes n}\,\hat{P}_0|x\rangle \\
&= -\hat{H}^{\otimes n}|x\rangle \\
&= -|\phi\rangle.
\end{aligned}
$$

Let us call

$$\hat{P}_\psi = \hat{H}^{\otimes n}\,\hat{P}_0\,\hat{H}^{\otimes n},$$

such that

$$
\begin{aligned}
\hat{P}_\psi|\psi\rangle &= |\psi\rangle \\
\hat{P}_\psi|\phi\rangle &= -|\phi\rangle
\end{aligned}
$$

for all $|\phi\rangle^{(n)}$. We can then rewrite the Grover iterate as

$$\hat{G} = \hat{P}_\psi\,\hat{U}_f.$$

Third, let us now go back and split the $n$-qubit state vector into the solution plus the non-solution parts, namely,

$$|\psi\rangle = \frac{1}{\sqrt{2^n}}|\omega\rangle + \sqrt{\frac{2^n - 1}{2^n}}|\Omega\rangle,$$

where $\langle\omega|\Omega\rangle = 0$. The vectors $|\omega\rangle$ and $|\Omega\rangle$ define a two-dimensional subspace. For convenience, we associate their amplitudes to sine and cosine functions of the same angle, namely,

$$\sin(\theta) = \frac{1}{\sqrt{2^n}} \quad \text{and} \quad \cos(\theta) = \sqrt{\frac{2^n - 1}{2^n}},$$

where $\theta = \sin^{-1}(2^{-n/2})$. For convenience (and to save some algebra), we can then build another vector $|\bar{\psi}\rangle$ that, together with $|\psi\rangle$, also spans a two-dimensional subspace:

$$\begin{aligned} |\psi\rangle &= \sin(\theta)|\omega\rangle + \cos(\theta)|\Omega\rangle \\ |\bar{\psi}\rangle &= \cos(\theta)|\omega\rangle - \sin(\theta)|\Omega\rangle, \end{aligned}$$

with $\langle\psi|\bar{\psi}\rangle = 0$. Now, let us apply first the black-box operator $\hat{U}_f$,

$$\begin{aligned} \hat{U}_f|\psi\rangle &= -\sin(\theta)|\omega\rangle + \cos(\theta)|\Omega\rangle \\ &= \cos(2\theta)|\psi\rangle - \sin(2\theta)|\bar{\psi}\rangle, \end{aligned}$$

and then apply the modified phase shifter,

$$\begin{aligned} \hat{P}_\psi \hat{U}_f|\psi\rangle &= \cos(2\theta)|\psi\rangle + \sin(2\theta)|\bar{\psi}\rangle \\ &= \sin(3\theta)|\omega\rangle + \cos(3\theta)|\Omega\rangle \end{aligned}$$

(there is some algebra and trigonometric relations involved in this last step). Thus,

$$\hat{G}|\psi\rangle = \sin(3\theta)|\omega\rangle + \cos(3\theta)|\Omega\rangle.$$

Notice that the result is a vector quite similar to $|\psi\rangle$, but coefficients where the angle is multiplied by a factor of 3. Even though $\theta$ is small, an increase by a factor of 3 means that $\hat{G}|\psi\rangle$ is closer to the solution $|\omega\rangle$ than $|\psi\rangle$ was.

Repeating this process $k$ times, we arrive at

$$\begin{aligned} (\hat{G})^k|\psi\rangle &= \cos(2k\theta)|\psi\rangle + \sin(2k\theta)|\bar{\psi}\rangle \\ &= \sin[(2k+1)\theta]|\omega\rangle + \cos[(2k+1)\theta]|\Omega\rangle. \end{aligned} \tag{9.2}$$

The idea now is to get $\sin[(2k+1)\theta]$ as close as possible to 1, such that the state vector is nearly identical to the solution vector $|\omega\rangle$ and all other orthogonal contributions are filtered out. Then $\omega$ can be obtained by a direct (single shot) measurement of the state in the computational basis. This is illustrated in Fig. 9.2.

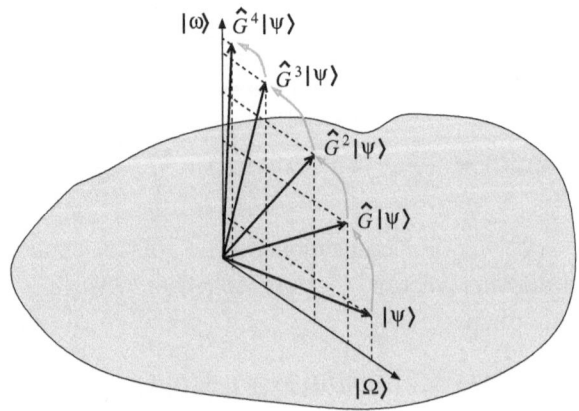

**Figure 9.2** Schematic illustration of the amplitude amplification process through the Grover iterate.

To get $\sin[(2k+1)\theta] \approx 1$ and $\cos[(2k+1)\theta] \approx 0$, we need to have $(2k+1)\theta \approx \pi/2$, which amounts to

$$k \approx \frac{\pi}{4\theta} - \frac{1}{2}$$
$$\approx \frac{\pi}{4}\sqrt{2^n}$$

for $n \gg 1$. Therefore, by applying the Grover iterate $\hat{G}$ on $|\psi\rangle^n$ $k$ times, where $k = \left\lfloor \frac{\pi}{4}\sqrt{2^n} - \frac{1}{2} \right\rfloor$, we can find the solution of $f(x) = 1$ with probability at least $O\left(1 - \frac{1}{N}\right)$, where $N = 2^n$. Therefore, we can obtain the solution with $O(1/N)$ accuracy by iterating $\hat{G}$ (i.e., calling the oracle) only $\sqrt{N}$ times. This is pretty good! We sped up the solution of the problem by a factor of $\sqrt{N}$ compared to the classical algorithm.

*A few comments*:

- The method of amplitude amplification used in Grover's search can be generalized to obtain the solution of any problem of the oracle type (namely, when it is efficient to verify whether a solution is correct or not).

- In the case of Grover's search, one builds a uniform superposition $n$-qubit states via a layer $\hat{H}^{\otimes n}$ of Hadamard gates, but other superposition states are also possible and can be tailored to the particular problem at hand.

- The amplitude amplification method can also be used to count (approximately) the number of solutions of a problem such as find $x$ such that $f(x) = 1$, as well as to prove if there are two distinct values $x$ and $y$ such that $f(x) = f(y)$.

## 9.2 APPLICATIONS

Because of its generality, Grover's search algorithm can be plugged into or adapted to various more specialized algorithms. Here, we list a few of them. They also target mathematical problems rather than specific practical challenges, but are very useful nevertheless.

### 9.2.1 Collision problem

Suppose you have black-box access to a multi-variable Boolean function $f : \{0,1\}^n \to \{1,0\}^n$, where $n$ is even. You are told that $f$ is a two-to-one functon: it takes exactly two different inputs to the same output. The collision problem amounts to finding $x$ and $y$ such that $x \neq y$ and $f(x) = f(y)$. It has an obvious application in cryptography: breaking hash functions. The best possible classical algorithm can solve this problem with $O(n^{1/2})$ queries, but Brassard, Høyer, and Tapp were able to apply Grover's search to this problem and solve it with $O(n^{1/3})$ queries.[2] Their algorithm works as follows:

- randomly select $n^{1/3}$ inputs, query them classically (i.e., compute their corresponding $f$ values), and sort them (e.g., in ascending order);

- run Grover's search algorithm over $n^{2/3}$ additional random inputs to $f$;

- select each input $x$ of the second query whenever $f(x) = f(y)$ for one of the $y$ inputs of the first query (this is where the sorting performed in the first step comes in handy).

Notice that the total number of queries to $f$ is $n^{1/3} + \sqrt{n^{2/3}} = O(n^{1/3})$.

---

[2]Brassard, G., P. Høyer, and A. Tapp. 1998. *Quantum cryptanalysis of hash and claw-free functions.* Springer, Lecture Notes in Computer Science, vol. 1380.

## 9.2.2  Quantum polling

Consider a string of $N$ bits. How many bits do you need to check in order the estimate the number of bits with value 1 within an error margin of $\pm\varepsilon N$? The classical way requires one to randomly and uniformly sample $O(1/\varepsilon^2)$ bits from the string and then take the average. Using Grover's search, it is possible to answer the question with just $O(1/\varepsilon)$ queries.

## 9.2.3  Quantum walks

There is an important class of algorithms named quantum walks that, while distinct from Grover's search, employs several elements from the latter. A quantum walk is defined on a graph, which is a collection of vertices connected by edges. The edges are typically directed, meaning they can only be taken in a given direction. In its classical version, the walker starts at a certain vertex; the walker then moves to an adjacent vertex specified by the result of a coin flip. The process continues in discrete time steps until the walker reaches the desired destination. Hence, random walks can be interpreted as a search on database with a graph structure.

In the quantum version, the coin is quantum and therefore superpositions of multiple adjacent vertices are allowed. Similarly to Grover's search, the quantum version offers a quadratic speed up with respect to the best classical algorithm: for a graph with $N$ nodes, a quantum walk algorithm requires only $O(\sqrt{N})$ discrete steps (i.e., calls to the oracle).[3]

Quantum random walks have been used as part of other quantum algorithms. For instance, consider the problem of computing the partition function of an interacting spin system (there are many important models of magnetic systems that fall into this category). Let $E(x)$ be the total energy of a configuration $x = (x_1, \ldots, x_N)$ of an $N$-spin system. The partition function is defined as weighed sum over all possible configurations,

$$Z(T) = \sum_x e^{-E(x)/k_B T},$$

where $T$ is the system's temperature. From this function, one can compute properties such as specific heat, which can be readily measured experimentally to confirm the accuracy of the model. The lower the temperature, the harder it becomes to even approximately compute $Z(T)$,

---

[3]Szegedy, M. 2004. *Quantum speed-up of Markov chain based algorithms*, Proceedings of the 45th Annual IEEE Symposium on Foundations of Computer Science: 32-41

as a larger number of contributions need to be taken into consideration. A combination of Grover's search, quantum walks, and phase estimation allows one to reduce the complexity of the calculation from $O(N^2/\delta\,\varepsilon^2)$ in the classical calculation to $O(N^2/\sqrt{\delta}\,\varepsilon)$, where $\delta$ is the energy difference between the two lowest energies the system can have and $\varepsilon$ is the desired accuracy of the computation.[4] The quantity $\delta$, the so-called spectral gap, is typically the most relevant scaling parameter, as its dependence on $N$ can vary a lot (from constant to exponential).

## 9.3 REFERENCES AND FURTHER READING

1. Mermin, N. D. 2007. *Quantum Computer Science*. Cambridge Univ. Press. Chapter 4.

2. Nielsen M. A. and I. L. Chuang. 2000. *Quantum Computation and Quantum Information*. Cambridge Univ. Press. Chapter 6.

3. Kaye, Ph., R. Laflamme and M. Mosca. 2007. *An Introduction to Quantum Computing*. Oxford Univ. Press. Chapter 8.

4. Williams, C. P. 2011. *Explorations in Quantum Computing*. Springer-Verlag. Chapter 5.

5. Aaronson, S. 2018. *Introduction to quantum information science – Lecture notes*. Chapter 24.

## 9.4 EXERCISES AND PROBLEMS

1. In the explanation of Grove's algorithm, we did not complete all steps in the derivations. You will do that here.

    (a) Show that
    $$\hat{H}^{\otimes n}\hat{P}_0^{\perp}\hat{H}^{\otimes n}|\psi\rangle^{(n)} = |\psi\rangle^{(n)}$$
    and
    $$\hat{H}^{\otimes n}\hat{P}_0^{\perp}\hat{H}^{\otimes n}|\phi\rangle^{(n)} = -|\phi\rangle^{(n)}.$$

    (b) Show that
    $$\hat{U}_f|\psi\rangle^{(n)} = \cos(2\theta)|\psi\rangle^{(n)} - \sin(2\theta)|\bar{\psi}\rangle^{(n)}$$

---

[4]Wocjan, P., C.-F. Chiang, D. Nagaj, and A. Abeyesinghe. 2009. *Quantum algorithm for approximating partition functions*. Phys. Rev. A 80:022340

and

$$\hat{H}^{\otimes n}\hat{P}_0^{\perp}\hat{H}^{\otimes n}\hat{U}_f|\psi\rangle^{(n)} = \sin(3\theta)|\omega\rangle^{(n)} + \cos(3\theta)|\Omega\rangle^{(n)}.$$

(Recall your trigonometric relations!)

2. Derive Eq. (9.2).

3. Build a "black-box" (oracle) operator $\hat{U}_f$ that implements the relation

$$\hat{U}_f|x\rangle^{(2)} \otimes |yw\rangle = |x\rangle^{(2)} \otimes |f(x) \oplus y\rangle,$$

where $|x\rangle^{(2)}$ represents a two-qubit computational basis state and $|y\rangle$ represents a single-qubit state, also in the computational basis. For this problem, the two-bit function $f(x)$ is defined as

$$f(x) = \begin{cases} 1, & x = 2 \\ 0, & x \neq 2 \end{cases}.$$

(Notice: 2 in decimal is equal to 10 in binary.) *Hint*: Find the unitary operator $\hat{U}_f$ in the three-qubit computational basis. What does it do?

# CHAPTER 10

# Density Matrices and Their Uses

So far we have dealt with quantum systems in a well-defined state, namely, systems that can be fully described by a single state vector $|\psi\rangle$. Such states are called *pure*. Here we will learn how to describe situations where knowing the state vector is not enough because the system is under the influence of another, typically larger, system which we know little about. This kind of interaction degrades the "quantumness" of the system we are interested to describe. The mathematics used to described this phenomenon is also suitable for establishing the degree of entanglement present in a quantum system, being it in a pure state or not.

## 10.1 MIXED STATES

Pure states are not the most general in quantum mechanics. In some situations, we cannot describe the state of a quantum system using a single state vector. This is typically the case when the system interacts with other systems or with its environment in a way that we cannot predict or control exactly. In these cases, we need to use a probabilistic approach to describe the state of the system. For instance, by associating probabilities to all possible state vectors that the system may have:

$$\{(p_1, |\psi_1\rangle), (p_2, |\psi_2\rangle), \ldots, (p_m, |\psi_m\rangle)\}.$$

Here, $p_k$ is the probability associated to a possible state vector $|\psi_k\rangle$, with $0 \leq p_k \leq 1$ and $k = 1, \ldots, m$.

This description is called an ensemble or a mixture of states. One refers to it as a *mixed* state, for short.

When you want to compute the expectation value of an operator $\hat{O}$ for a system in a mixed state, you resort to

$$\langle \hat{O} \rangle = \sum_{k=1}^{m} p_k \langle \psi_k | \hat{O} | \psi_k \rangle,$$

i.e., you perform an ensemble average of all the expectation values of the same operator for all vector states that they system may have, weighed by their respective probabilities.

## 10.2   DENSITY MATRICES

Working with probabilities can be cumbersome; there is actually an equivalent but more compact way to handle mixed states that takes the probabilities into account without explicitly writing them down. It is called the density matrix formulation. Density matrices are defined as

$$\hat{\rho} = \sum_{k=1}^{m} p_k | \psi_k \rangle \langle \psi_k |. \tag{10.1}$$

Since $\sum_{k=1}^{m} p_k = 1$, the trace of this matrix must be 1:

$$
\begin{aligned}
\mathrm{Tr}[\hat{\rho}] &= \sum_{n} \langle n | \hat{\rho} | n \rangle \\
&= \sum_{n} \sum_{k=1}^{m} p_k \langle n | \psi_k \rangle \langle \psi_k | n \rangle \\
&= \sum_{k=1}^{m} p_k \sum_{n} |\langle n | \psi_k \rangle|^2 \\
&= \sum_{k=1}^{m} p_k \\
&= 1.
\end{aligned}
$$

(Here, we made use of a complete basis $\{|n\rangle\langle n|\}$ and assumed that the states $\{|\psi_k\rangle\}$ are normalized.)

It is straightforward to check that $\hat{\rho}$ is Hermitian, namely, $\hat{\rho}^{\dagger} = \hat{\rho}$.

For an $n$-qubit system, we have $1 \leq m \leq 2^n$. Because the states $|\psi_k\rangle$ are $n$-dimensional, the matrix $\hat{\rho}$ has dimensions $2^n \times 2^n$. Another important property of density matrices is that they are semi-positive: they can only have zero or positive eigenvalues. This is manifest by the semi-positive matrix elements along the main diagonal.

Notice that a pure state is representable by a single state vector, say $|\psi\rangle$, since $m = 1$. Hence, the corresponding density matrix is the outer product

$$\hat{\rho} = |\psi\rangle\langle\psi|.$$

Moreover, $\hat{\rho} = \hat{\rho}^2$ in this case (the density matrix of a pure state is idempotent).

It is straightforward to compute expectation values with density matrices: for an operator $\hat{O}$, we have

$$\langle\hat{O}\rangle = \text{Tr}\left[\hat{\rho}\hat{O}\right].$$

Since any quantum state (pure or mixed) can be represented by a density matrix, given a density matrix, how do we know whether the state is pure or mixed? To answer this question, notice that, from Eq. (10.1), we can write

$$
\begin{aligned}
\hat{\rho}^2 &= \sum_{k=1}^{m}\sum_{k'=1}^{m} p_k\, p_{k'}\, |\psi_k\rangle\langle\psi_k||\psi_{k'}\rangle\langle\psi_{k'}| \\
&= \sum_{k=1}^{m}\sum_{k'=1}^{m} p_k\, p_{k'}\, |\psi_k\rangle\, \delta_{k,k'}\, \langle\psi_{k'}| \\
&= \sum_{k=1}^{m} p_k^2\, |\psi_k\rangle\langle\psi_k|.
\end{aligned}
$$

Therefore,

$$\text{Tr}\left[\hat{\rho}^2\right] = \sum_{k=1}^{m} p_k^2.$$

Since $\sum_{k=1}^{m} p_k = 1$, using the triangular inequality it follows that

$$\text{Tr}\left[\hat{\rho}^2\right] \leq 1.$$

The equal sign is only possible if the state is pure, namely, when $m = 1$. In summary,

$$\text{Tr}\left[\hat{\rho}^2\right] = 1 \quad \Rightarrow \quad \text{pure state}$$

$$\text{Tr}\left[\hat{\rho}^2\right] < 1 \quad \Rightarrow \quad \text{mixed state.}$$

There is another, somewhat more laborious way to determine if the state is pure or mixed: if the eigenvalues of $\hat{\rho}$ are all either 0 or 1, the state is pure.

## 10.2.1 Single-qubit mixed states

We have seen that pure single-qubit states can be visualized as points on the surface of the Bloch sphere. What about mixed states? They correspond to points *inside* the Bloch sphere! In fact, a maximally mixed state corresponds to a point at the center of the sphere.

(A maximally mixed state in the context of single-qubit states corresponds to the ensemble $\{(\frac{1}{2}, |0\rangle), (\frac{1}{2}, |1\rangle)\}$.)

We will not go into all the mathematical details here, but the connection between the single-qubit density matrix and the Bloch sphere can be established as follows:

1. Since $\hat{\rho}$ is a $2 \times 2$ matrix, it can decomposed in terms of Pauli matrices, namely,

$$\hat{\rho} = \frac{1}{2}\left[\hat{I} + a_x\,\hat{X} + a_y\,\hat{Y} + a_z\,\hat{Z}\right].$$

2. The three-dimensional vector defined by the amplitudes $\vec{a} = (a_x, a_y, a_z)$ is called the Bloch vector.

3. The polar and azimuth angles can be obtained through the standard spherical coordinate decomposition:

$$
\begin{aligned}
a_x &= |\vec{a}|\sin\theta\,\cos\phi \\
a_y &= |\vec{a}|\sin\theta\,\sin\phi \\
a_z &= |\vec{a}|\cos\theta.
\end{aligned}
$$

4. Because $\mathrm{Tr}\left[\hat{\rho}^2\right] \le 1$, it follows that $a_x^2 + a_y^2 + a_z^2 \le 1$ as well. For a pure state, $|\vec{a}| = 1$ and the tip of the Bloch vector is on the surface of the sphere. For mixed states, $|\vec{a}| < 1$ and the tip of the vector is inside the sphere.

## 10.2.2 Reduced density matrices

Consider a system of two qubits in a pure product state,

$$|\psi_{AB}\rangle = |\psi_A\rangle \otimes |\psi_B\rangle.$$

The corresponding density operator is

$$
\begin{aligned}
\hat{\rho}_{AB} &= |\psi_{AB}\rangle\langle\psi_{AB}| \\
&= |\psi_A\rangle\langle\psi_A| \otimes |\psi_B\rangle\langle\psi_B|.
\end{aligned}
$$

Since the state is pure, we must have (check it!)

$$\text{Tr}\left[\hat{\rho}_{AB}^2\right] = 1. \tag{10.2}$$

The trace here is taken over all basis states, including degrees of freedom (i.e., variables) of both qubits. We can define a *partial trace* that sums over only the degrees of freedom of one qubit:[1]

$$\text{Tr}_B[\hat{\rho}_{AB}] = |\psi_A\rangle\langle\psi_A| \times \text{Tr}[|\psi_B\rangle\langle\psi_B|]$$
$$\text{Tr}_A[\hat{\rho}_{AB}] = \text{Tr}[|\psi_A\rangle\langle\psi_A|] \times |\psi_B\rangle\langle\psi_B|.$$

We associate to each partial trace a *reduced density matrix*,

$$\hat{\rho}_A = \text{Tr}_B\left[\hat{\rho}_{AB}\right] \quad \text{and} \quad \hat{\rho}_B = \text{Tr}_A\left[\hat{\rho}_{AB}\right].$$

The partial trace amounts to a Hilbert space reduction. If the original density matrix $\hat{\rho}_{AB}$ had dimensions $2^n \times 2^n$ and $n = n_A + n_B$, then the reduced density matrices $\hat{\rho}_A$ and $\hat{\rho}_B$ will have dimensions $2^{n_A} \times 2^{n_A}$ and $2^{n_B} \times 2^{n_B}$, respectively. Essentially, we are summing over one qubit while the other qubit stays as a spectator.

Because we started with a product state, notice that

$$\hat{\rho}_A = |\psi_A\rangle\langle\psi_A| \quad \text{and} \quad \hat{\rho}_B = |\psi_B\rangle\langle\psi_B|$$

since $\text{Tr}[|\psi_A\rangle\langle\psi_A|] = 1$ and $\text{Tr}[|\psi_B\rangle\langle\psi_B||] = 1$ as well. Therefore,

$$\text{Tr}\left[\hat{\rho}_A^2\right] = 1 \quad \text{and} \quad \text{Tr}\left[\hat{\rho}_B^2\right] = 1.$$

Both qubits $A$ and $B$ are in pure states after the partial traces. But what if we had started with an entangled state? For example, the state vector

$$|\psi_{AB}\rangle = \frac{1}{\sqrt{2}}(|0\rangle_A \otimes |1\rangle_B + |1\rangle_A \otimes |0\rangle_B) = \frac{1}{\sqrt{2}}(|01\rangle + |10\rangle),$$

which yields the density matrix

$$\hat{\rho}_{AB} = \frac{1}{2}(|01\rangle\langle01| + |01\rangle\langle10| + |10\rangle\langle01| + |10\rangle\langle10|).$$

---

[1]When the trace has no subscript, it is implicitly assumed that the sum is over all degrees of freedom, resulting in a scalar.

(Notice that $\text{Tr}\left[\hat{\rho}_{AB}^2\right] = 1$ since this is still a pure state.) Let us take a partial trace:

$$\begin{aligned}
\hat{\rho}_A &= \text{Tr}_B\left[\hat{\rho}_{AB}\right] \\
&= \frac{1}{2}\left(|0\rangle\langle0|\,\text{Tr}\left[|1\rangle\langle1|\right] + |0\rangle\langle1|\,\text{Tr}\left[|1\rangle\langle0|\right] + |1\rangle\langle0|\,\text{Tr}\left[|0\rangle\langle1|\right]\right. \\
&\quad \left. + |1\rangle\langle1|\,\text{Tr}\left[|0\rangle\langle0|\right]\right).
\end{aligned}$$

Since

$$\text{Tr}\left[|0\rangle\langle0|\right] = \text{Tr}\left[|1\rangle\langle1|\right] = 1 \quad \text{and} \quad \text{Tr}\left[|0\rangle\langle1|\right] = \text{Tr}\left[|1\rangle\langle0|\right] = 0,$$

we obtain

$$\hat{\rho}_A = \frac{1}{2}\left(|0\rangle\langle0| + |1\rangle\langle1|\right),$$

which describes a single-qubit in a mixed state. Let us check:

$$\begin{aligned}
\text{Tr}\left[\hat{\rho}_A^2\right] &= \frac{1}{4}\text{Tr}\left[\left(|0\rangle\langle0| + |1\rangle\langle1|\right)\left(|0\rangle\langle0| + |1\rangle\langle1|\right)\right] \\
&= \frac{1}{2}\text{Tr}\left[|0\rangle\langle0| + |1\rangle\langle1|\right] \\
&= \frac{1}{2}.
\end{aligned}$$

This result provides an insight into the inner workings of quantum systems:

1. The partial trace operation can be thought of as a way to impose our ignorance about one of the subsystems. For instance, suppose that our focus is only subsystem $A$ and thus we trace $B$ out. We will then run into one of two possible outcomes:

    (a) a pure state for subsystem $A$ if $A$ and $B$ were not entangled (e.g., because they had not previously interacted with each other);

    (b) a mixed state for subsystem $A$ if $A$ and $B$ were entangled.

    If subsystem $A$ is in a mixed state, it means that $A$ was entangled to subsystem $B$.

2. It is interesting to revisit two-qubit protocols such as teleportation in light of these findings. One can prove that that if Alice and Bob share an entangled state, nothing that Alice does to her qubits affects Bob's *reduced* density matrix. Thus there is no instantaneous information propagation. Quantum mechanics and special relativity are compatible!

### 10.2.3 State purification

Given a mixed state, is it possible to construct a pure state by enlarging the system? The answer is positive and the process is called purification. Here is how it works.

1. Let

$$\hat{\rho}_A = \sum_{k=1}^{m} p_k |\psi_k\rangle_A \langle_A \psi_k|$$

   represent a mixed state in the space $H_A$.

2. Find the eigenvalues and eigenvectors of $\hat{\rho}_A$,

$$\hat{\rho}_A |\phi_j\rangle = \lambda_j |\phi_j\rangle, \tag{10.3}$$

   with $j = 1, \ldots, 2^{n_A}$.

3. Select $m$ eigenvalues and eigenvectors of $\hat{\rho}_A$. Combine them with the states of another system, $B$ which is a copy of $A$, as in

$$|\psi_{AB}\rangle = \sum_{k=1}^{m} \sqrt{p_k} \, |\psi_k\rangle_A \otimes |\phi_k\rangle_B.$$

4. Now, $|\psi_{AB}\rangle$ is a pure state and

$$\mathrm{Tr}_B \left[|\psi_{AB}\rangle\langle\psi_{AB}|\right] = \hat{\rho}_A.$$

### 10.2.4 Equation of motion for density matrices

Let us recall the equation of motion that describes the time evolution of a ket state vector (i.e., Schrödinger's equation):

$$\frac{d}{dt}|\psi\rangle = \frac{-i}{\hbar} \hat{H}|\psi\rangle,$$

where $\hat{H}$ is the Hamiltonian operator of the system. Similarly, for the bra version of the state vector,

$$\frac{d}{dt}\langle\psi| = \frac{i}{\hbar} \langle\psi|\hat{H}.$$

We can employ these expressions to derive an equation of motion for the density matrix. We start with

$$\hat{\rho} = \sum_k p_k |\psi_k\rangle\langle\psi_k|.$$

Taking the time derivative, we obtain[2]

$$
\begin{aligned}
\frac{d}{dt}\hat{\rho} &= \sum_k p_k \left[ \left( \frac{d}{dt}|\psi_k\rangle \right) \langle\psi_k| + |\psi_k\rangle \left( \frac{d}{dt}\langle\psi_k| \right) \right] \\
&= \frac{i}{\hbar} \sum_k p_k [(-\hat{H}|\psi_k\rangle)\langle\psi_k| + |\psi_k\rangle(\langle\psi_k|\hat{H})] \\
&= -\frac{i}{\hbar} \sum_k p_k [\hat{H}, |\psi_k\rangle\langle\psi_k|] \\
&= \frac{1}{i\hbar}[\hat{H}, \hat{\rho}],
\end{aligned}
$$

where we introduced the commutator $[\hat{A}, \hat{B}] = \hat{A}\hat{B} - \hat{B}\hat{A}$. This first-order differential equation is fully equivalent to Schrödinger's equation, but applicable to density matrices.

Recall that there is another approach to time evolution which resorts to the use of operators. For state vectors, we have

$$
|\psi(t)\rangle = \hat{U}(t,0)|\psi(0)\rangle.
$$

When a similar approach is applied to density matrices, we find

$$
\begin{aligned}
\hat{\rho}(t) &= \hat{U}(t,0)\,\hat{\rho}(0)\,[\hat{U}(t,0)]^\dagger \\
&= \hat{U}(t,0)\,\hat{\rho}(0)\,\hat{U}(0,t).
\end{aligned}
$$

## 10.3  DECOHERENCE

We now show how to extend the equation of motion to include interactions with an environment, which leads the system to lose its quantum identity via a process called decoherence. When the information from the quantum system is rapidly dissipated within the environment, it is possible to disregard memory effects and derive a simply local-in-time correction to the density matrix equation-of-motion in an weak-coupling approximation:

$$
\frac{d}{dt}\hat{\rho} = \frac{1}{i\hbar}[\hat{H}, \hat{\rho}] + \sum_k \left( 2\,\hat{L}_k\hat{\rho}\hat{L}_k^\dagger - \{\hat{L}_k^\dagger\hat{L}_k, \hat{\rho}\} \right), \tag{10.4}
$$

where $\{\cdots\}$ denotes an anticommutator: $\{\hat{A}, \hat{B}\} = \hat{A}\hat{B} + \hat{B}\hat{A}$. This expression is called the Lindblad equation and $\hat{L}_k$ are called Lindblad operators.[3] They are judiciously chosen, on a case-by-case basis, depending

---

[2]For those more versed in quantum mechanics: do not confound this result with Heisenberg's equation of motion for operators!

[3]Lindblad operators do not need to be Hermitian.

on the relevant system-environment interactions. An important aspect of the Lindblad equation is that it preserves key properties of the density matrix, such as unit trace and hermiticity.

Let us put it to use. Consider a qubit system with no interactions other than the one with an environment that favors a transition from the state $|1\rangle$ to the state $|0\rangle$,[4] which can be captured by the Lindblad operator

$$\hat{L} = \sqrt{\lambda}\,|0\rangle\langle 1| = \begin{pmatrix} 0 & 1 \\ 0 & 0 \end{pmatrix},$$

where $\lambda$ is a positive coupling constant. Denoting the density matrix as

$$\hat{\rho} = \begin{pmatrix} \rho_{00} & \rho_{01} \\ \rho_{10} & \rho_{11} \end{pmatrix},$$

after some algebra, we obtain the following Lindblad equation:

$$\frac{d}{dt}\begin{pmatrix} \rho_{00} & \rho_{01} \\ \rho_{10} & \rho_{11} \end{pmatrix} = \lambda \begin{pmatrix} 2\rho_{11} & -\rho_{01} \\ -\rho_{10} & -2\rho_{11} \end{pmatrix}. \tag{10.5}$$

This matrix equation represents a system of four first-order linear differential equations (two of them coupled) on the variables $\rho_{00}$, $\rho_{01}$, $\rho_{10}$, and $\rho_{11}$. Its general solution is derived in Appendix A.4 and results in

$$\begin{pmatrix} \rho_{00}(t) & \rho_{01}(t) \\ \rho_{10}(t) & \rho_{11}(t) \end{pmatrix} = \begin{pmatrix} 1 - \rho_{11}(0)\,e^{-2\lambda t} & \rho_{01}(0)\,e^{-\lambda t} \\ \rho_{10}(0)\,e^{-\lambda t} & \rho_{11}(0)\,e^{-2\lambda t} \end{pmatrix}. \tag{10.6}$$

Let us assume that the system starts in the pure superposition state

$$|\psi(0)\rangle = \frac{1}{\sqrt{2}}\left(|0\rangle + |1\rangle\right),$$

which can be cast as a density matrix by setting $\rho_{00}(0) = \rho_{01}(0) = \rho_{10}(0) = \rho_{11}(0) = \frac{1}{2}$. Inserting these values in Eq. (10.6), we find

$$\hat{\rho}(t) = \frac{1}{2}\begin{pmatrix} 1 - e^{-2\lambda t} & e^{-\lambda t} \\ e^{-\lambda t} & e^{-2\lambda t} \end{pmatrix}.$$

Notice that this density matrix describes mixed state since $\mathrm{Tr}\left[\hat{\rho}^2\right] < 1$. It is clear that for times larger $1/\lambda$, the system decays toward the $|0\rangle$ state.

---

[4]For example, if $|0\rangle$ and $|1\rangle$ are low- and high-energy electronic states of an atom, this interaction would correspond to the emission of a photon.

The characteristic time associated to the decay toward a (classical) basis state is called relaxation time and denoted as $T_1$. It can be retrieved by inspecting the time dependence of the diagonal matrix elements. In this example, $T_1 = 1/(2\lambda)$.

Notice that the off-diagonal matrix elements also decay, but twice as slow, with a characteristic time $T_2 = 1/\lambda$, which is called dephasing time. It is possible to have dephasing without relaxation. Consider another type of system-environment interaction where the Lindblad operator is given by[5]

$$\hat{L} = \sqrt{\lambda}\,\hat{Z},\tag{10.7}$$

implying a Lindblad equation of the form

$$\frac{d}{dt}\begin{pmatrix} \rho_{00} & \rho_{01} \\ \rho_{10} & \rho_{11} \end{pmatrix} = -2\lambda \begin{pmatrix} 0 & \rho_{01} \\ \rho_{10} & 0 \end{pmatrix}.\tag{10.8}$$

It is clear that now the diagonal matrix elements are constant while the off-diagonal ones decay with a characteristic time $T_2 = 1/(2\lambda)$.

## 10.4 ENTANGLEMENT ENTROPY

We can quantify the amount of mixing present in a mixed state, and by the same token the amount of entanglement in an entangled state. The method to do so goes back to Claude Shannon, a pioneer of classical information theory who was interested in quantifying information content in the early days of telecommunications.[6]

Suppose there is a message $m$ with length $l$ (e.g, the number of bits needed to express the message). Among all possible messages of length $l$, let $p_m$ be the probability of $m$. Shannon defined the information content of the message $m$ as

$$I_m = -\log_2(p_m).$$

The smaller the chances of the message $m$ occur, the higher its information content.

Since $I_m$ is itself a stochastic (i.e., random) variable, we can define its ensemble average as

$$H \equiv \overline{I_m} = -\sum_m p_m \log_2(p_m).$$

---

[5]This interaction corresponds to the environment randomly flipping the relative phase of the qubit.

[6]Shannon, C. E. 1948. *A mathematical theory of communications*. Bell System Technical Journal 27: 379-423, 623-656

$H$ is called Shannon information entropy.

We can apply the same concept to a mixed state, namely, $\{(p_1, |\psi_1\rangle), (p_2, |\psi_2\rangle), \ldots\}$, in which case we define

$$S = -\sum_k p_k \ln(p_k).$$

In the context of quantum mechanics, $S$ is known as the von Neumann entropy. In terms of density matrices, we find that

$$S = -\operatorname{Tr}[\hat{\rho}\ln(\hat{\rho})].$$

(Notice the natural logarithm as opposed to the logarithm base 2 of Shannon entropy.)

Recall that we can express the logarithm of an operator as a Taylor series expansion. For instance,

$$\ln(\hat{\rho}) = (\hat{\rho} - \hat{I}) - \frac{1}{2}(\hat{\rho} - \hat{I})^2 + \frac{1}{3}(\hat{\rho} - \hat{I})^3 - \cdots$$

when expanded around the identity operator. After a bit of algebra, we find

$$\hat{\rho}\ln(\hat{\rho}) = \hat{\rho}(\hat{\rho} - \hat{I})\left[\hat{I} - \frac{1}{2}(\hat{\rho} - \hat{I}) + \frac{1}{3}(\hat{\rho} - \hat{I})^2 - \cdots\right]$$

$$= (\hat{\rho}^2 - \hat{\rho})\left[\hat{I} - \frac{1}{2}(\hat{\rho} - \hat{I}) + \frac{1}{3}(\hat{\rho} - \hat{I})^2 - \cdots\right].$$

Thus, for a pure state, since $\hat{\rho} = \hat{\rho}^2$, we find $S = 0$. A pure state has zero von Neumann entropy. A mixed state, however, will have a finite von Neumann entropy. Therefore, computing $S$ is another way to distinguish between pure and mixed states.

But there is more we can do with the von Neumann entropy.

Going back to entangled states, consider a system formed by two subsystems $A$ and $B$. We can quantify the amount of entanglement between $A$ and $B$ through the entropy associated to the reduced density matrix:

$$S_A = -\operatorname{Tr}[\hat{\rho}_A \ln \hat{\rho}_A],$$

where $\hat{\rho}_A = \operatorname{Tr}_B[\hat{\rho}_{AB}]$. Similarly,

$$S_B = -\operatorname{Tr}[\hat{\rho}_B \ln \hat{\rho}_B]],$$

where $\hat{\rho}_B = \operatorname{Tr}_A[\hat{\rho}_{AB}]$. When $\hat{\rho}_{AB}$ represents a pure stay, $S_A = S_B$. The higher the value of $S_A$ (or $S_B$), the more entanglement there is between $A$ and $B$. The von Neumann entropy is upper bounded by the dimension of the Hilbert space, namely,

$$0 \le S \le \ln(d),$$

where $d = 2^n$ for a system of $n$ qubits.

There are many other ways to quantify entanglement. Interestingly, they are all equivalent to each other in some way, and are monotonic functions of each other (although findings these functional relations can be extremely difficult).

Yet, from a computational perspective, calculating these measures of entanglement is always hard, typically exponentially hard, requiring $O(2^n)$ operations when using a classical computer. In fact, it has been rigorously proved that deciding whether a given $\hat{\rho}_{AB}$ matrix represents a product or an entangled state is an NP-hard problem, in general.

There are many other interesting properties related to entanglement entropies, but they are beyond the scope of this book.

## 10.5 REFERENCES AND FURTHER READING

1. Schumacher B. and M. Westmoreland. 2010. *Quantum Processes, Systems, and Information*. Cambridge Univ. Press. Sections 8.1-8.3.

2. Liboff, R. L. 2003. *Introductory Quantum Mechanics*, 4th edition. Addison Wesley. Section 11.11.

3. Nielsen M. A. and I. L. Chuang. 2000. *Quantum Computation and Quantum Information*. Cambridge Univ. Press. Sections 2.4, 2.5, and 8.3

4. Williams, C. P. 2011. *Explorations in Quantum Computing*. Springer-Verlag. Sections 11.2, 11.3, and 14.1.

## 10.6 EXERCISES AND PROBLEMS

1. Verify that Eq. (10.2) is valid.

2. Consider the two-qubit entangled pair in the teleportation protocol. Show that the reduced density matrix for Bob's qubit is independent of the state of Alice's qubit.

3. Prove that when a system composed of two subsystems $A$ and $B$ is in a pure state, the entanglement entropies of those subsystems are equal, namely, $S_A = S_B$.

4. Derive Eq. (10.5) starting from the general Lindblad equation (10.4))and the choice of Lindblad operator shown in Eq. (10.3).

5. Similarly, derive Eq. (10.8).

6. Show that Eq. (10.6) represents a solution of Eq. (10.5). If you feel ambitious, you can employ the method of Appendix A.4, although in this particular case substantial simplifications are possible.

# CHAPTER 11

# Quantum Error Correction

All physical systems that provide a good template for the implementation of qubits and quantum gates suffer from some degree of decoherence. The reason is simple: no physical system exists in complete isolation; if it did exist, it would be useless for quantum computing since you would not be able to interact with it! Any physical system, even when well protected, will eventually interact with its environment in a way that we cannot predict or completely control. As a result, coherence, which is essential for quantum information processing, is progressively lost. In the field of quantum computation, it was realized early on that strategies must be employed to mitigate not only imprecisions in the instrumentation but also errors associated to decoherence and relaxation due to interactions with the environment. These strategies, when employed at the quantum circuit level, are called quantum error correction (QEC).

Errors do not afflict only quantum information processing. They are also common in classical information processing, from computing to communications. Very clever error correction methods have been devised to deal with such errors, and QEC borrows heavily from those methods. Typically, one introduces extra, "redundant" bits and extra encoding to build "logical bits" which employ several physical bits to represent, say, a single binary state. The same strategy is adopted in QEC, but there are some important differences. Firstly, in classical computing, errors amount to bit flips whereas for qubits one also needs to worry about "phase" flips. As a result, QEC employs more complex encoding than classical error correction. Secondly, we cannot copy or clone qubit states as we do for bits.

The encoding in QEC must be such that one can not only detect errors when they occur but also correct for them. Since one cannot clone quantum states, the task of detecting and recovering is far from trivial. Moreover, the fact that qubits can accumulate errors continuously and that any attempt to measure a qubit irreversibly collapses its quantum state further complicate the problem.

For those reasons, many people thought that quantum computers, despite their amazing powers (e.g., they can implement Shor's factoring algorithm), were completely impractical. But Peter Shor came to the rescue: in 1995, soon after proposing his famous quantum algorithm, he also showed that a robust, single-qubit QEC code existed (it utilizes 9 physical qubits in total).[1] In addition, he showed how to quantum compute directly on logical spaces. These developments, as well as similar work done by others, brought back the hope that quantum computing was actually possible. Since then, many other people have contributed to this area and even better quantum error correction schemes have been proposed. Nowadays, there is little doubt that QEC will play a major role in any large-scale implementation of quantum information processing.

QEC is itself a vast and well-developed subfield of QIP. Let us look at some basic QEC strategies.

## 11.1 DEALING WITH QUBIT-FLIP ERRORS

A simple way to detect bit flip errors (which are classical errors) is via a majority rule: instead of using only one physical qubit to store $|0\rangle$ and $|1\rangle$ states, we can use three qubits, as in $|000\rangle$ and $|111\rangle$, and find out through some kind of measurement if one of these physical qubits has flipped (we assume that enough time has elapsed between preparation and measurement such that a flip might have happened due to the interaction with the environment). In the case that a flip was detected and identified, we can then proceed to correct it.

This is easier said than done because of the no-cloning theorem, but here is a strategy. Let us start with the encoding.

1. Consider a single-qubit in a generic (unknown) state

$$|\psi\rangle = \alpha|0\rangle + \beta|1\rangle.$$

2. Initialize two ancillary qubits to $|0\rangle$.

---

[1]Shor, P. W. 1995. *Scheme for reducing decoherence in quantum computer memory*. Phys. Rev. A 52: R2492-R2496

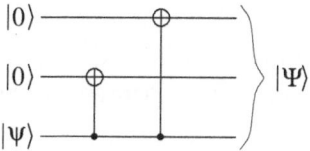

Figure 11.1  Encoding for a three-qubit logical qubit.

3. Apply two CNOT gates in succession, using the qubit on the $|\psi\rangle$ state as control and setting targets on the ancillary qubits, as shown in Fig. 11.1, resulting in the three-qubit state

$$|\Psi\rangle = |0\rangle \otimes |0\rangle \otimes |\psi\rangle = |0\rangle \otimes |0\rangle \otimes (\alpha|0\rangle + \beta|1\rangle) = \alpha|000\rangle + \beta|111\rangle.$$

We have basically encoded the single-qubit state into a higher-dimensional Hilbert space. Any single-qubit flip will rotate the state vector $|\Psi\rangle$, but only within a well-defined subspace of the higher-dimensional Hilbert space. How do we detect such a rotation? The circuit in Fig. 11.2 is an example which requires adding another ancillary qubit (recall that a control-$X$ gate is the same as a CNOT gate).

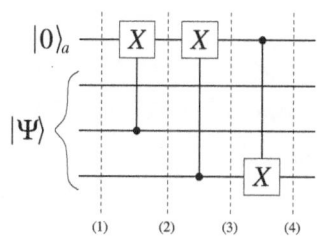

Figure 11.2  QEC circuit for a qubit flip.

The circuit in Fig. 11.2 detects and corrects for a flip on the first qubit in $|\Psi\rangle$ (we enumerate the qubits from the bottom). Let us see how this works.

First, notice that if the state $|0\rangle \otimes |000\rangle$ was input into this circuit, it would go through it without any alteration (check it!). However, if the state $|0\rangle \otimes |111\rangle$ was input instead it would transform twice and then be reconstituted at the end of the circuit (check that too!).

This means that the two "logical" states

$$|0\rangle_L \equiv |000\rangle$$
$$|1\rangle_L \equiv |111\rangle$$

are not affected by the circuit. What about a state with a bit flip error, say one where the third qubit has been flipped? This would have been caused by an error that can be represented by the operator

$$\hat{E} = \hat{I} \otimes \hat{I} \otimes \hat{X}.$$

In fact,

$$\hat{E}(\alpha|000\rangle + \beta|111\rangle) = \alpha|001\rangle + \beta|110\rangle.$$

Let us check step-by-step the evolution of the states $|001\rangle$ and $|110\rangle$ (which represent a bit flip on the third qubit) as they go through the circuit. We start with $|001\rangle$:

$$
\begin{array}{lll}
@\,\text{stage1}: & |0\rangle_a \otimes |001\rangle \\
@\,\text{stage2}: & |0\rangle_a \otimes |001\rangle \\
@\,\text{stage3}: & |1\rangle_a \otimes |101\rangle \\
@\,\text{stage4}: & |1\rangle_a \otimes |000\rangle.
\end{array}
$$

Notice that the flip error on the third qubit was corrected. A measurement on the ancillary qubit will reveal the occurrence of the error. Now let us consider $|110\rangle$:

$$
\begin{array}{lll}
@\,\text{stage1}: & |0\rangle_a \otimes |110\rangle \\
@\,\text{stage2}: & |1\rangle_a \otimes |110\rangle \\
@\,\text{stage3}: & |1\rangle_a \otimes |110\rangle \\
@\,\text{stage4}: & |1\rangle_a \otimes |111\rangle.
\end{array}
$$

Similarly, the error was corrected and a measurement of the ancillary qubit will reveal the error occurrence without affecting the state of the logical qubit.

Notice, however, that if the bit flip error had occured in any of the other two qubits (i.e, the first or the second), the circuit would fail to detect and correct for that error.

We can generalize this circuit to one where qubit flip errors in any physical qubit of the logical qubit can be corrected, see Fig. 11.3. Notice that two ancillary qubits are necessary in this case.

There is an alternative way to perform the syndrome part (i.e., the error detection part) which is more suitable in practice and more useful for correcting other types of errors. It relies on the circuit identity shown in Fig. 11.4. The result is shown in Fig. 11.5.

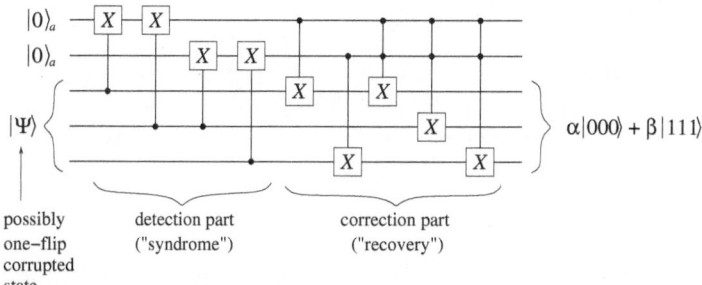

Figure 11.3 Circuit to correct for a qubit flip in any physical qubit of the logical qubit.

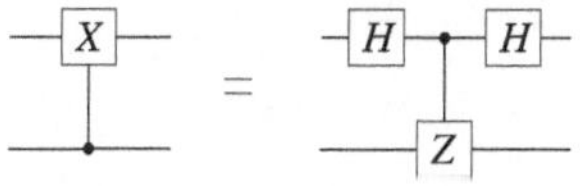

Figure 11.4 CNOT decomposition in terms of Hadamard and Ctrl-Z gates.

Formally, we have found a way to keep the codewords $|000\rangle$ and $|111\rangle$ protected against the following error operators:

$$\hat{E}_0 = \hat{I} \otimes \hat{I} \otimes \hat{I}$$
$$\hat{E}_1 = \hat{X} \otimes \hat{I} \otimes \hat{I}$$
$$\hat{E}_2 = \hat{I} \otimes \hat{X} \otimes \hat{I}$$
$$\hat{E}_3 = \hat{I} \otimes \hat{I} \otimes \hat{X}.$$

Calling $\hat{U}_{s+r}$ the unitary operator for the complete error correction circuit (syndrome + recovery parts combined), we have

$$\hat{U}_{s+r} \cdot \left( |00\rangle_a \otimes \hat{E}_k |\Psi\rangle \right) = |\phi\rangle_a \otimes |\Psi\rangle,$$

where $k = 0, 1, 2, 3$. Here, $|00\rangle_a$ and $|\phi\rangle_a$ stand for the ancillary qubits on input and on output, respectively.
*Note*: in terms of the density matrices defined in Chapter 10, we have

$$\hat{\rho}_{\text{in}} = |00\rangle_{aa}\langle 00| \otimes |\Psi\rangle\langle\Psi| \quad \text{(product state)}$$

$$\implies \quad \text{tr}_a(\hat{\rho}_{\text{in}}) = |\Psi\rangle\langle\Psi|$$

$$\hat{\rho}_{\text{in}} + \text{error} = |00\rangle_{aa}\langle 00| \otimes \hat{E}_k|\Psi\rangle\langle\Psi|\hat{E}_k^\dagger \quad \text{(product state)}$$

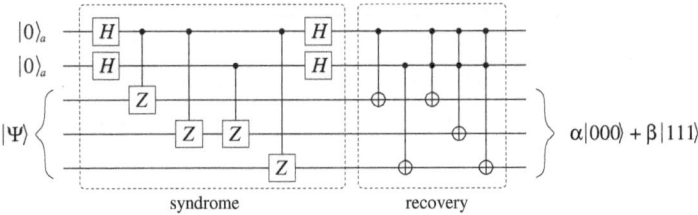

Figure 11.5 Same as in Fig. 11.3 but incorporating the circuit identity of Fig. 11.4. Notice that we have reverted back to the standard notation for CNOT and Toffoli gates.

$$
\begin{aligned}
\hat{\rho}_{\text{out}} &= \hat{U}_{s+r} \cdot \hat{\rho}_{\text{in}} + \text{err} \cdot \hat{U}_{s+r}^{\dagger} \\
&= \hat{U}_{s+r} \cdot (|00\rangle_{aa}\langle 00| \otimes \hat{E}_k |\Psi\rangle\langle\Psi| \hat{E}_k^{\dagger}) \cdot \hat{U}_{s+r}^{\dagger} \\
&= \hat{U}_{s+r} \cdot (|00\rangle_a \hat{E}_k |\Psi\rangle\langle\Psi| \hat{E}_{k a}^{\dagger}\langle 00|) \cdot \hat{U}_{s+r}^{\dagger} \\
&= |\phi\rangle_a \otimes |\Psi\rangle\langle\Psi| \otimes {}_a\langle\phi|
\end{aligned}
$$

$$
\implies \quad \text{tr}_a(\hat{\rho}_{\text{out}}) = |\Psi\rangle\langle\Psi|.
$$

*Very important note*: the circuit only corrects for a single qubit-flip error. If two qubit-flips occur simultaneously, they will not be properly detected and corrected. More ancillary qubits and a more complex circuit would be needed to properly handle such errors. However, on physical grounds one expects double errors to be much less probable than single errors: if qubit-flip errors are statistically independent and uncorrelated, then the probability of a double error relates to the probability of a single error as

$$
p_{\text{double}} = p_{\text{single}}^2.
$$

Therefore, we can conclude that

$$
p_{\text{double}} \ll p_{\text{single}}
$$

so long as $p_{\text{single}} \ll 1$, which is a reasonable assumption (else, we picked a very bad quantum system to begin with!).

## 11.2  PHASE-FLIP ERRORS

This type of error occurs in quantum systems only and amounts to a spurious change in the internal relative phase of a quantum state. For instance,

$$
|\psi\rangle = \alpha|0\rangle + \beta|1\rangle \longrightarrow |\psi'\rangle = \alpha|0\rangle - \beta|1\rangle.
$$

In this particular example, the phase-flip error can be represented by the action of the $\hat{Z}$ operator. (The phase-flip error is very much analogous to the ordinary flip error, but in a rotated basis.) We will adopt the $\hat{Z}$ operator to describe phase-flip errors; other descriptions are possible upon a suitable change of basis states. Notice that

$$\hat{Z}|+\rangle = |-\rangle \quad \text{and} \quad \hat{Z}|-\rangle = |+\rangle,$$

where

$$|\pm\rangle = \frac{1}{\sqrt{2}}(|0\rangle \pm |1\rangle).$$

Thus, we need to encode $|\psi\rangle$ into an entangled state involving $|+\rangle$ and $|-\rangle$ single-qubit states in order to handle phase-flip errors. For instance,

$$\alpha|0\rangle + \beta|1\rangle \longrightarrow \alpha|+++\rangle + \beta|---\rangle.$$

The way to do it is via a layer of Hadamard gates, as shown in Fig. 11.6.

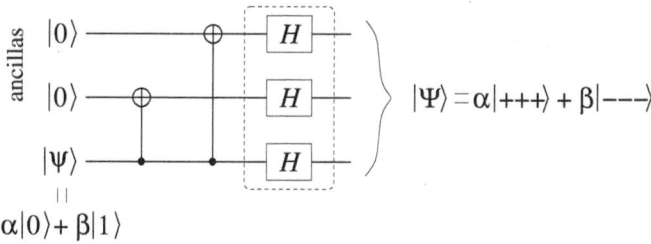

Figure 11.6 Enconding for phase-flip error correction.

Except for the addition of Hadamard gates, the error correction circuit for phase-flip errors is similar to that used for bit-flip errors, see Fig. 11.7.

To understand the role of the layers of Hadamard gates before the syndrome and after the recovery stages, notice that

$$\hat{H}|+\rangle = |0\rangle \quad \text{and} \quad \hat{H}|-\rangle = |1\rangle.$$

The Hadamard gates essentially rotate the qubit states to a basis where errors can be detected and corrected like bit flips, and then return the states to the original basis.

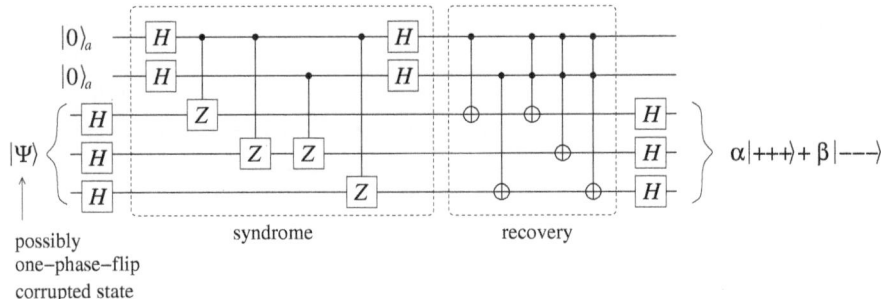

Figure 11.7 Circuit for phase-flip error correction.

## 11.3 MORE GENERAL ERRORS

We have taken care of bit-flip and phase-flip errors. What about more general errors? And what if we do not know which form of error is prevalent in the qubit system we are interested?

Shor's 1996 QEC code takes care of the most general single-qubit error possible, which is of the form

$$\hat{E} = a_I \hat{I} + a_x \hat{X} + a_y \hat{Y} + a_z \hat{Z},$$

where $a_k$ is some amplitude coefficient, with $k = I, x, y, z$. (Notice that $\hat{I}$ is not an error but is included to cover the case when no error has taken place, which is always a possibility.) The code requires 9 physics qubits. The codewords are

$$|0\rangle_L = \frac{1}{\sqrt{8}} (|000\rangle + |111\rangle) \otimes (|000\rangle + |111\rangle) \otimes (|000\rangle + |111\rangle)$$

$$|1\rangle_L = \frac{1}{\sqrt{8}} (|000\rangle - |111\rangle) \otimes (|000\rangle - |111\rangle) \otimes (|000\rangle - |111\rangle),$$

or, equivalently but more compactly,

$$|0\rangle_L = |+++\rangle \otimes |+++\rangle \otimes |+++\rangle$$
$$|1\rangle_L = |---\rangle \otimes |---\rangle \otimes |---\rangle.$$

(Make sure you understand that these are nine-qubit states.) The encoding, syndrome, and recovery circuits appropriate for this encoding are generalizations of those used for the bit-flip and phase-flip errors; the methodology is the same. The circuits are too large to show here. The reader is asked to check the end-of-chapter references for the diagrammatic representations of those circuits.

There are other classes of quantum error correction codes beyond Shor's, including some very clever ones. For instance, Andrew Steane has showed that using the encoding

$$|0\rangle_L = \frac{1}{\sqrt{8}}(|0000000\rangle + |1010101\rangle + |0110011\rangle + |1100110\rangle)$$
$$+ |0001111 + |1011010\rangle + |0111100\rangle + |1101001\rangle)$$
$$|1\rangle_L = \frac{1}{\sqrt{8}}(|1111111\rangle + |0101010\rangle + |1001100\rangle + |0011001\rangle)$$
$$+ |1110000 + |0100101\rangle + |1000011\rangle + |0010110\rangle),$$

which involves "only" seven physical qubits, it is possible to correct for any single-qubit error, just like in Shor's encoding. Others have shown that five-qubit quantum codes that correct for any error do exist. It seems that no such a code with less than five qubits exist.

It is understandable that, for newcomers, the whole QEC business looks opaque, and more art than science. But it turns out that there is a theoretical framework backing it up and it goes by the name of stabilized theory.[2] It is a beautiful but somewhat complicated theory created by Daniel Gottesman. Here we will only cover its fundamentals, so that you can get a general idea about how it works.

## 11.4 STABILIZER THEORY

Consider the two-qubit state (which happens to be a Bell state ):

$$|\Psi\rangle = \frac{1}{\sqrt{2}}(|0\rangle_1 \otimes |0\rangle_2 + |1\rangle_1 \otimes |1\rangle_2).$$

Notice that

$$\hat{X}_1\hat{X}_2|\Psi\rangle = |\Psi\rangle$$

and

$$\hat{Z}_1\hat{Z}_2|\Psi\rangle = |\Psi\rangle.$$

We say that $|\Psi\rangle$ is stabilized by the operators $\hat{M}_1 = \hat{X}_1\hat{X}_2$ and $\hat{M}_2 = \hat{Z}_1\hat{Z}_2$ since it is invariant under the action of these operators (a.k.a. stabilizers). Actually, it turns out that $|\Psi\rangle$ is uniquely defined by its

---

[2]Gottesman, D. 1997. *Stabilizer codes and quantum error correction*. Ph.D. thesis, California Institute of Technology, Pasadena, CA

stabilizers, up to a global phase. The idea of the stabilizer theory is to work with the stabilizers of a state instead of the state itself. This has many advantages, among them the ability to unify all different quantum error correction codes in a more concise description (a family of codes is described by their common stabilizer set).

Let us go back to Shor's nine-qubit code, which has codewords

$$
\begin{aligned}
|0\rangle_L &= |+++\rangle \otimes |+++\rangle \otimes |+++\rangle \\
|1\rangle_L &= |---\rangle \otimes |---\rangle \otimes |---\rangle.
\end{aligned}
$$

One can check that the set of operators

$$
\begin{aligned}
\hat{M}_1 &= \hat{X}_6 \hat{X}_5 \hat{X}_4 \hat{X}_3 \hat{X}_2 \hat{X}_1 \\
\hat{M}_2 &= \hat{X}_9 \hat{X}_8 \hat{X}_7 \hat{X}_6 \hat{X}_5 \hat{X}_4 \\
\hat{M}_3 &= \hat{Z}_2 \hat{Z}_1 \\
\hat{M}_4 &= \hat{Z}_3 \hat{Z}_2 \\
\hat{M}_5 &= \hat{Z}_5 \hat{Z}_4 \\
\hat{M}_6 &= \hat{Z}_6 \hat{Z}_5 \\
\hat{M}_7 &= \hat{Z}_8 \hat{Z}_7 \\
\hat{M}_8 &= \hat{Z}_9 \hat{Z}_8
\end{aligned}
$$

stabilize the logical states $|0\rangle_L$ and $|1\rangle_L$, namely,

$$
\begin{aligned}
\hat{M}_j |0\rangle_L &= |0\rangle_L \\
\hat{M}_j |1\rangle_L &= |1\rangle_L
\end{aligned}
$$

for $j = 1, \ldots, 8$. Moreover,

$$
[\hat{M}_j, \hat{M}_{j'}] = 0
$$

for all $j, j'$, i.e., the stabilizers commute amongst themselves.

These operators are members of a larger set called the Pauli group. (A group is basically a set that closes in itself upon an operation such as multiplication.) The size of the Pauli group depends on the number of qubits involved in the operations. The single-qubit Pauli group is given by[3]

$$
P_1 = \left\{ \pm\hat{I}, \pm i\hat{I}, \pm\hat{X}, \pm i\hat{X}, \pm\hat{Y}, \pm i\hat{Y}, \pm\hat{Z}, \pm i\hat{Z} \right\}.
$$

---

[3]For 9 qubits, the Pauli group comprises $4 \times 4^9 = 4^{10} = 20^{20} = (2^{10})^2 = (1024)^2 \approx 1,000,000$ elements. Therefore, it is too large to be enumerated here! But the nine-qubit code stabilizer group is much smaller.

Notice that any product of two elements results in another element of the same set, and that is why we call this set a group. A stabilizer group is a subgroup of the Pauli group for the given number of qubits where two conditions must be satisfied: the $-\hat{I}$ operator is excluded and all elements commute.

The eight operators $\hat{M}_j$ are special because they span (i.e., generate) all elements of the stabilizer group of 9 qubits. Notice that they themselves do not form a subgroup.

Each codeword set (and Shor's encoding being a particular case) has its own stabilizer group.

The most interesting feature of the stabilizer theory is that any operator in the Pauli group $P_n$ that is not part of a stabilizer group $S$ anticommutes with the elements of $S$. Therefore, any error that can be cast as an operator outside a stabilizer group $S$ can be detected and corrected by measuring the operators in $S$. Here is a protocol for that:

1. if all measurements of operators in $S$ yield $+1$, no error is detected and no recovery is needed;

2. if a subset of measurements yield $-1$, then an error has occurred and must be corrected; the specific recovery operation varies depending on the set of $\pm 1$ values measured, as each type of single-qubit error produces a different signature.

To understand point 2, suppose that $\hat{O}$ represents an error that is not part of the space $S$ and let $\hat{M}_i$ be one of the stabilizers which anticommutes with $\hat{O}$. Let $|\psi'\rangle = \hat{O}|\psi\rangle$ be the affected state. Then,

$$
\begin{aligned}
\langle \psi' | \hat{M}_i | \psi' \rangle &= \langle \psi | \hat{O}^\dagger \hat{M}_i \hat{O} | \psi \rangle \\
&= -\langle \psi | \hat{O}^\dagger \hat{O} \hat{M}_i | \psi \rangle \\
&= -\langle \psi | \psi \rangle \\
&= -1.
\end{aligned}
$$

Measurments of the stabilizer $\hat{M}_i$ will yield $-1$, indicating that an error of type $\hat{O}$ has occurred. The particular set of stabilizers that yield $-1$ upon measurements constitutes a "fingerprint" of the error.

## 11.5  FAULT-TOLERANT QUANTUM COMPUTING

QEC provides a framework for a more robust (but costly!) way to process quantum information, often referred as fault tolerant quantum computing. The basic idea is to compute on logical states (which are encoded with multiple physical qubits) rather than on individual physical qubits. For instance, to use logical Pauli gates such as in

$$\hat{X}_L|0\rangle_L = |1\rangle_L$$
$$\hat{X}_L|1\rangle_L = |0\rangle_L.$$

The logical operator $\hat{X}_L$ likely involves multiple operations on the physical qubits encoding the logical states. The hope is that this approach will be less prone to errors at the logical level than when doing computations directly with physical qubits.[4] There is actually a rigorous proof that this strategy works, so long as one can hierarchically build more and more levels of logical qubits and gates, depending on how high the single-qubit error probability is. The original concept dates back to von Neumann, one of the pioneers of modern classical computing and is illustrated by the diagram in Fig. 11.8:

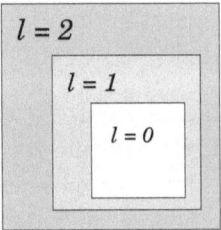

Figure 11.8  Schematic representation of fault-tolerant quantum error correction.

$l = 0$: physical level
$l = 1$: first logical level
$l = 2$: second logical level
and so on. The process is called concatenation. For instance, using as an example Shor's nine-qubit code,

$$|0\rangle, |1\rangle \;:\; l = 0$$
$$|0\rangle_L, |1\rangle_L \;:\; l = 1$$
$$|0\rangle_{L^2}, |1\rangle_{L^2} \;:\; l = 2$$
$$\vdots \qquad \vdots$$

[4]Shor, P. 1996. *Fault-tolerant quantum computation.* Proceedings of 37th Conference on Foundations of Computer Science: 56–65

where

$$|0\rangle_L = |+++\rangle|+++\rangle|+++\rangle \quad \text{(9 physical qubits involved)}$$
$$|1\rangle_L = |---\rangle|---\rangle|---\rangle$$

$$|0\rangle_{L^2} = |+++\rangle_L|+++\rangle_L|+++\rangle_L \quad \text{(81 physical qubits involved)}$$
$$|1\rangle_{L^2} = |---\rangle_L|---\rangle_L|---\rangle_L.$$

Therefore, the amount of resources needed for fault-tolerant quantum computing increases rapidly (exponentially) with the hierarchy level. However, the reduction in the probability of uncorrected errors decreases even faster: if $p$ is the error probability at level 0 ($p < 1$), the error probability at level $l$ goes as $p^{2^l}$ (an exponential of an exponential in $l$). For a fixed $p$, the net result is that the amount of resources needed scales polynomially with the number of gates and qubits at level 0, whenever a fixed error rate $\epsilon$ per operation is required or acceptable. For instance, when the goal is $p_l < \epsilon$, it is straightforward to show that $l < \log_2(\frac{\log \epsilon}{\log p})$, which indicates very shallow hierarchical encodings may be enough.

This result is called the threshold theorem and is more rigorously stated as: provided that $p < p^*$, to achieve a fixed single-qubit error rate $\epsilon$ one only needs a polynomial amount of resources to implement fault tolerance.[5]

This looks very good! But there is a caveat: the theorem is based on the (often implicit) hypothesis that errors at the physical level ($l = 0$) are uncorrelated in time and space. There are known deviations from this theorem when this assumption is not satisfied. Solid-state qubits in particular are prone to correlated errors.

*Notes:*

- One often speaks of stabilizer gates (CNOT, Hadamard, $P$), as well as of stabilizer circuits, which are circuits made entirely of stabilizer gates. It turns out that stabilizer circuits fall into the class of Clifford circuits and can be efficiently simulated by classical computers (Gottesman-Knill theorem). This is the reason why IBM Quantum offers a simulator called "stabilizer", with thousands of qubits. These are not real qubits, but classically simulated ones. The reason why one can reach such a large number of "qubits" is that Clifford gates do not form a universal set and therefore

---

[5]The actual value of $p^*$ varies from one type of encoding to another.

cannot generate intricate entangled quantum states, the types of which are required for quantum algorithms that offer exponential speed up (like Shor's).

- In practice, non-stabilizer gates are much harder to implement than stabilizer ones!

## 11.6 REFERENCES AND FURTHER READING

1. Mermin, N. D. 2007. *Quantum Computer Science.* Cambridge Univ. Press. Chapter 5.

2. Nielsen M. A. and I. L. Chuang. 2000. *Quantum Computation and Quantum Information.* Cambridge Univ. Press. Chapter 10 and Section 11.3.

3. Kaye, Ph., R. Laflamme and M. Mosca. 2007. *An Introduction to Quantum Computing.* Oxford Univ. Press. Chapter 10.

4. Williams, C. P. 2011. *Explorations in Quantum Computing.* Springer-Verlag. Sections 14.3-14.10.

## 11.7 EXERCISES AND PROBLEMS

1. Show that the circuit in Fig. 11.3 corrects for flip errors in any of the physical qubits comprising the logical qubit.

2. Prove the circuit identity shown in Fig. 11.4.

3. Consider a three-qubit quantum code $|0\rangle_L = |000\rangle$ and $|1\rangle_L = |111\rangle$ that protects against qubit flip errors.

    (a) Define a circuit to perform encoding for this quantum error correction code, namely, a circuit that takes $|\psi\rangle = \alpha|0\rangle + \beta|1\rangle$ into $|\psi\rangle_L = \alpha|0\rangle_L + \beta|1\rangle_L$.

    (b) Show that if, at most, one one-qubit flip error occurs on the codeword, after decoding using the inverse of the encoding circuit, a Toffoli gate will recover the original state $|\psi\rangle$ on the first qubit and the error will be transferred to ancillary qubits.

# CHAPTER 12

# Alternative Forms of Quantum Computing

The approach we considered so far for computing using quantum mechanics relies on preparing a suitable initial state, applying single and two-qubit quantum gates, and performing measurements. It is known as the circuit model of quantum computing. Howeveer, this is not the only approach. In this chapter, we will cover a few alternatives. Some still rely on circuits to some extend, but they either explore alternative paradigms of fault tolerance, utilize protocols that do not require entangling gates, or avoid multi-shot computations. One of these alternative approaches is entirely based on measurements. Another requires fine tunning gates and repeating measurments.

## 12.1 ADIABATIC QUANTUM COMPUTING

The fundamental physical principle behind this method is rather old (it goes back to the pioneering years of quantum mechanics), but it was only first applied to quantum computing in 2000.[1]

Let $\hat{H}_1$ be a Hamiltonian operator such that its ground state encodes the solution of a computation problem of interest. For example, imagine that we want to find the value of $x \in \{0,1\}^n$ such that $f(x)$ is minimal. This function could be the sum of pairwise interactions, such as

$$f(x) = \sum_{i \neq j} a_{ij} \left( x_i - \frac{1}{2} \right) \left( x_j - \frac{1}{2} \right),$$

---

[1]Farhi, E., J. Goldstone, S. Gutman, M. Sipser. 2000. *Quantum computation by adiabatic evolution.* arXiv:quant-ph/0001106

where $a_{ij}$ are random coefficients quantifying the interaction between sites $i$ and $j$ on a lattice. (This particular functional form applies to a large class of combinatorial optimization problems.) $\hat{H}_1$ is usually entirely classical and its ground state is also classical (i.e., a product state). Yet, finding its exact ground state can be a very hard problem to solve.

Now, imagine that there is another Hamiltonian, $\hat{H}_0$, which has a very simple, easy-to-prepare ground state. The idea of the method is to prepare qubits in the ground state of $\hat{H}_0$ and act on these qubits with a Hamiltonian that continuously and slowly deforms $\hat{H}_0$ into $\hat{H}_1$, so that when the deformation is completed, the qubits find themselves in the ground state of $\hat{H}_1$, thus encoding the desired solution. In other words, the method drives the qubit state from the "easy" ground state to the "complicated" one. At the end of the process, one can retrieve the solution by just measuring the state of the qubits in the computational basis. (Since, at this point, the state should be classical, the measurements yield the solution with 100% certainty.)

Formally, let the total (time-dependent) Hamiltonian be defined as

$$\hat{H}(t) = \left(1 - \frac{t}{T}\right)\hat{H}_0 + \left(\frac{t}{T}\right)\hat{H}_1,$$

where $T$ is some suitable time interval. Notice that

$$\hat{H}(0) = \hat{H}_0 \quad \text{and} \quad \hat{H}(T) = \hat{H}_1.$$

Therefore, $\hat{H}(t)$ continuously interpolates between the two extreme Hamiltonians.

The hope is that by adopting a large-enough $T$, the system will remain in the instantaneous ground state of $\hat{H}(t)$ at all times $0 \leq t \leq T$. In fact, there is a well-known theorem in quantum mechanics that assures success of this strategy when the following condition is satisfied: if

$$\frac{\hbar \left| \left\langle \frac{d\hat{H}}{dt} \right\rangle_{1,0} \right|}{\Delta_{\min}^2} \leq \epsilon, \tag{12.1}$$

then

$$|\langle \psi_0(t) | \psi_0^{(1)} \rangle|^2 \geq 1 - \epsilon^2, \tag{12.2}$$

where

- $|\psi_0(t)\rangle$ is the instantaneous ground state of $\hat{H}(t)$ with corresponding energy $E_0(t)$;

- $|\psi_1^{(1)}\rangle$ is the ground state of $\hat{H}_1$;

- $\Delta_{\min} = \min[E_1(t) - E_0(t)]_{0 \leq t \leq T}$ is the smallest energy gap between ground and first excited state during the evolution (see Fig. 12.1);

- $|\psi_1(t)\rangle$ is the first excited state of $\hat{H}(t)$ with energy $E_1(t)$;

- $\left\langle \frac{d\hat{H}}{dt} \right\rangle_{1,0} = \langle \psi_1(t) | \frac{d\hat{H}}{dt} | \psi_0(t) \rangle$ is the matrix element of the derivative of the $\hat{H}(t)$ between ground state and first-excited state.

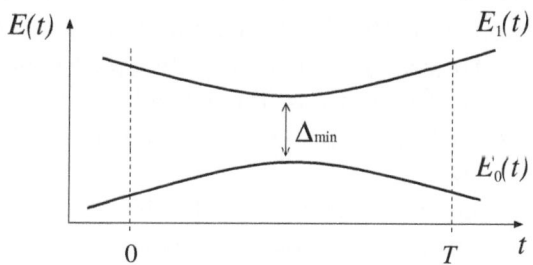

Figure 12.1 Schematic representation of the gap between the ground state and first-excited state during an adiabatic evolution.

Notice that $\left\langle \frac{d\hat{H}}{dt} \right\rangle_{1,0}$ measures the rate of change of the Hamiltonian. The larger the rate (i.e., the shorter the $T$), the harder it becomes to satisfy condition (12.1). That is why the method is called "adiabatic" (in reference to slow processes in thermodynamics, where there is no heat exchange with the environment). Together, the inequalities (12.1) and (12.2) are known as the adiabatic theorem. When $T$ is too large, the system can jump into a superposition between ground and excited states, scrambling the measurements and corrupting the solution.

The adiabatic approach has been proved to be equivalent to the circuit model of quantum computing. Both approaches lead to universal quantum computing. The main question is whether the adiabatic method provides a quantum speed up in comparison to classical computing. The answer lies on how $\Delta_{\min}$ scales with the number of qubits $n$. Since

$$\left\langle \frac{d\hat{H}}{dt} \right\rangle_{1,0} \approx \frac{E_c}{T},$$

where $E_c$ is some characteristic energy of the system (presumably weakly dependent on size or number of qubits), then, for a fixed $\epsilon$,

$$\frac{\hbar E_c}{T} \leq \Delta_{\min}^2 \, \epsilon \quad \Longrightarrow \quad T \geq \frac{1}{\epsilon} \frac{E_c}{\Delta_{\min}^2}.$$

When $\Delta_{\min} \sim O(1/n^p)$ for some positive power $p$, $T \sim O(n^{2p})$, which implies polynomial complexity. Good! But when $\Delta_{\min} \sim O(e^{-n})$, then $T \sim O(e^{2n})$, i.e., exponential complexity. Not good!

There is a large literature showing that for most problems of practical interest, $\Delta_{\min}$ either vanishes (really bad!) or becomes exponentially small somewhere between $t = 0$ and $t = T$. The physical reason for the vanishing gap is that the Hamiltonian undergoes a first-order phase transition for any trajectory in the parameter space connecting $\hat{H}_0$ and $\hat{H}_1$. For these kinds of transitions, the ground and first excited energy levels either cross each other or become extremely close, leading to an exponential slow down.

There exist some encodings that try to circumvent first-order transitions but usually other issues pop up, such as super-slow relaxations.

Nevertheless, these difficulties have not prevented people from building hardware to implement adiabatic quantum computing. A pioneer in this business is D-Wave in Canada. Their latest processors contain several thousands of qubits. But much still remains to be studied in this area, including the role of error correction and approximate solutions. For instance, approximate solutions to optmization problems obtained via adiabatic quantum computing could be used in conjunction with machine learning or be a component of it.

## 12.2  MEASUREMENT-BASED QUANTUM COMPUTING

In the circuit-based model of quantum computing, it is essential to be able to perform at least one type of two-qubit entangling operation, such as a CNOT, in addition to one-qubit operations. But high-fidelity two-qubit operations are usually hard to implement in hardware and on-demand (in particular when they involve photonic qubits), hence the question arises whether it would be possible to do universal quantum computation without them. The answer is yes, but a shift in paradigm is needed. The alternative is based on the preparation of special types of entangled states before the actual computation. The computation then proceeds by suitable measurements of these states using a particular basis, as well as the application of single-qubit gates. Basically, the idea is to start with a many-qubit state that contains all the resources needed to perform the computation, and then just steer the system toward the target state which contains the result of the computation. And the steering is performed through measurements and single-qubit operations only. It is no surprise that the underlying foundation of this method is teleportation.

Here is how it works.

Start with two single-qubit states $|\psi\rangle$ and $|\phi\rangle$ and two Bell-state pairs of the type $|B_{00}\rangle$. See Fig. 12.2, which comprises two teleportation circuits side-by-side.

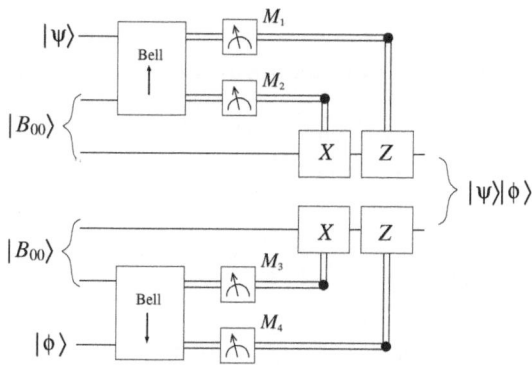

Figure 12.2  Building block circuit for measurement-based quantum computing.

The Bell boxes in Fig. 12.2 correspond to the inverse Bell-state preparation circuits used in the teleportation protocol, as shown in Fig. 12.3.

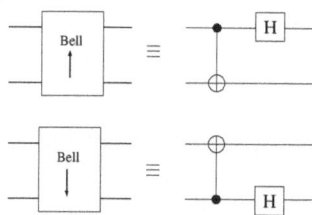

Figure 12.3  Bell state boxes.

We do not know what the states $|\psi\rangle$ and $|\phi\rangle$ are, but we are certain that, on output, the operations described in the double teleportation circuit diagram result in a two-qubit product state $|\psi\rangle \otimes |\phi\rangle$.

If we want to apply a CNOT gate on this two-qubit product state, we add such a gate to the output side and incorporate it into the measurements. But how? Here is a way: we can add two CNOT gates between the middle bitlines (since CNOT gates are their own inverse, applying two of them in sequence has no net effect). See Fig. 12.4.

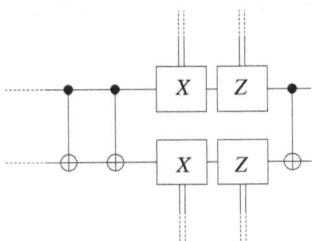

Figure 12.4 CNOT insertions for the measurement-based circuit.

We can then incorporate one of the CNOTs on the left into the measurements and use the other CNOT to modify the Bell states, as shown in Fig. 12.5.

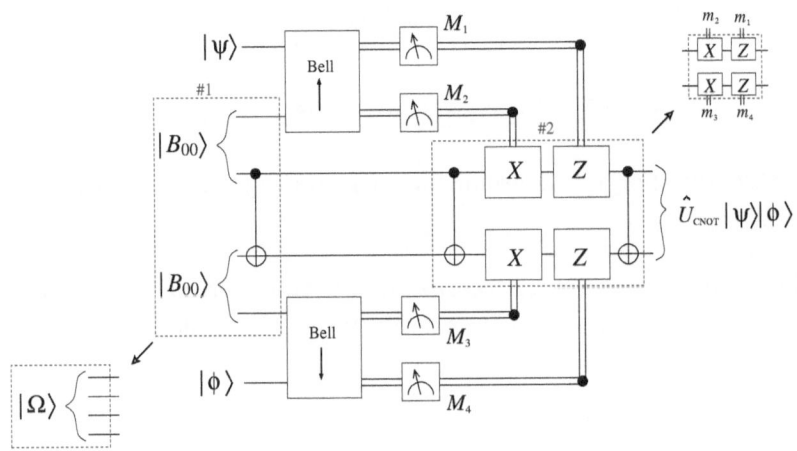

Figure 12.5 Implementation of measurement-based CNOT gate.

Notice the two dashed-line boxes, #1 and #2. Box #1 describes a new, four-qubit entangled state,

$$|\Omega\rangle = \frac{1}{2}(|0000\rangle + |0011\rangle + |1101\rangle + |1110\rangle).$$

$|\Omega\rangle$ is a special four-qubit entangled state that can be prepared in advance, stored in memory, and used on demand, namely, every time one needs to perform a CNOT operation on the single-qubit states $|\psi\rangle$ and $|\phi\rangle$.

Box #2 is equivalent to two original (disconnected) sequences of $\hat{X}$ and $\hat{Z}$ operators, but controlled by four integers $m_1$, $m_2$, $m_3$, and $m_4$ instead of the original $M_1$, $M_2$, $M_3$, and $M_4$. In other words, it is possible to convert $\{M_1, M_2, M_3, M_4\}$ into $\{m_1, m_2, m_3, m_4\}$ in such a way that the two CNOT operators in that box are absorbed.

We then conclude that, provided we can:

1. prepare the four-qubit state $|\Omega\rangle$

2. make measurements directly in the Bell basis and apply single-qubit gates,

we can avoid explicitly applying a CNOT gate when we need to perform such an operation on two single-qubit states $|\psi\rangle$ and $|\phi\rangle$.

Of course, it is not always obvious that avoiding entangling gates is a good tradeoff, given the additional preparation overhead and qubits needed. But for optical realizations of quantum computing, this approach was a tremendous boost when invented by D. Gottesman and I. Chuang in 1999.[2] The reason is because soon after, in 2001, E. Knill, R. Laflamme, and G. Milburn showed that it could be used to implement universal quantum computing with only linear optical elements and photodetectors (a method coined LOQC for linear optics quantum computing).[3] Since then, it has been experimentally demonstrated. The nonlinearity needed for quantum computing is embedded in the state $|\Omega\rangle$ and in the measurements. This method is referred as teleportation-based quantum computing.[4]

It turns out that one can extend this method further and drop single-qubit gates as well; it is possible to run the computation entirely through measurements. This is possible so long as a special type of multi-qubit entangled states, known as cluster states, can be prepared and utilized as initial states of the quantum processor. Essentially, cluster states contain all the resources needed to achieve the desired computation, with measurements being used just to steer the state toward the desired outcome. This approach is called "one-way quantum computing" and was

---

[2]Gottesman, D. and I. L. Chuang. 1999. *Demonstrating the viability of universal quantum computation using teleportation and single-qubit operations*. Nature 402: 390-393

[3]Knill, E., R. Laflamme, and G. J. Milburn. 2001. *A scheme for efficient quantum computation with linear optics*. Nature 409: 46-52

[4]Here, for simplicity, we have omitted an important fact: the two-qubit entangling gate in LOQC, a control-$Z$ gate, is actually probabilistic; to increase its success rate to practical levels, a quantum teleportation version of it is employed.

invented by Robert Raussendorf and Hans Briegel in 2001.[5] This method has also been proved to achieve universal quantum computing. However, the preparation of the cluster states can be extremely challenging. The method is also popular in the linear optics community, as well as in the area of optical lattices.

One-way quantum computing can also be used in conjunction with another form of quantum computing called topological (see Sec. 12.3).

There have not been many attempts to implement one-way quantum computing experimentally. In addition to being difficult to prepare cluster states, measuring individually (often closely-packed) qubits is hard.

## 12.3  TOPOLOGICAL QUANTUM COMPUTING

We have seen that quantum error correction, combined with fault-tolerant approaches, can reduce the effects of noise on qubits. However, these mitigation methods introduce a large hardware overhead, not to mention complex operations.

In 1997, Alexei Kitaev came up with the idea of encoding quantum information (i.e., qubits) in topological degrees of freedom that are immune to local errors.[6] The fundamental physics behind this idea is that environmental noise acts locally, thus if the logical qubit states are encoded in a non-local way, they becomes resilient to noise up to a very high degree. Topological states offer such nonlocality because they involve a multitude of physical degrees of freedom yet can be defined by just a few indices or numbers. A topological state can be "deformed", but as long as their characteristic indices do not change, the state does not change either.

The standard example of topological invariance is a torus of genus 1 (i.e., a three-dimensional object with a single piercing hole). A donut falls into this category, but so does a coffee mug with a single handle: in principle, one can continuously deform a coffee mug into a donut and vice versa without changing is topology (i.e., the genus). See Fig. 12.6. By the same reasoning, a solid sphere and solid cube are topologically indistinguishable (both have genus 0).

There are electronic and spin states in condensed matter systems that mimic the donut-mug scenario, namely, they can be reshaped without

---

[5]Raussendorf, R. and H. J. Briegel. 2001. *A one-way quantum computer*. Phys. Rev. Lett. 86: 5188-5191

[6]Kitaev, A. Yu. 2003. *Fault-tolerant quantum computation by anyons*. Ann. Phys. (N.Y.) 303: 2-30

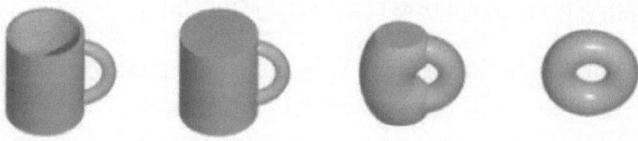

Figure 12.6  A mug can be made into a dognut by a continuous deformation.

losing their topological indices. If you just disturb the state a little and locally, without changing its topological nature, information encoded in the state is not lost. As a result, such states are extremely resilient to noise. In recent years, more and more of such states have been proposed theoretically; some have even been experimentally verified (still, theory is far ahead of experiments in this area as topological systems are very difficult to synthesize).

We would need an entire chapter (and a lot of physics) to review in detail the various systems proposed for topological quantum computing. Here is a partial list of the most prominent candidates with succinct explanations:

- non-Abelian excitations with fractional statistics (they appear in certain materials in the presence of strong magnetic fields);

- excitations that behave like a special type of fermion called Majorana (they are believed to exist in semiconductor nanowires capped with superconductor films);

- spin-1/2 systems on square lattices with four-body interactions, or where four-body measurements are possible (nature does not seem to provide such a system, so it would have to be built artificially);

- vortices in a class of superconductors where electronic states have some peculiar (topological) features (several candidate materials exist).

Not a single one of these proposed topological systems has produced any usable qubit yet. Still, hopes are high and a large number of researchers continue to study and investigate this topic. The challenges in materials synthesis and device fabrication are enormous.

## 12.4 VARIATIONAL QUANTUM COMPUTING

While we wait for high-fidelity, large-scale qubit systems to become a reality, we need to find ways to make use of noisy, intermediate-scale quantum (NISQ) hardware that is currently available. This hardware is not suitable for the implementation of complex and resource-hungry quantum algorithms like Shor's and Grover's, much less quantum error correction, but they can be utilized for tasks that do not require universal quantum computing capabilities. In particular, one can use present-day hardware, where the number of qubits ranges from a few tens to a few hundred, to solve for the ground state of physical systems relevant to materials science and chemistry. The reason is that finding accurately the ground state of certain compounds and molecules is extremely difficult because of the large Hilbert space that one needs to mimic when using a classical computer. But when one can directly translate electronic orbits into quantum mechanical degrees of freedom such as qubits, the calculation may be possible.

The methodology for this kind of quantum mechanical calculation is based on the variational theorem: let $\hat{H}$ be the Hamiltonian operator of the system of interest. For any trial state $|\psi\rangle$, the quantity

$$E = \frac{\langle \psi | \hat{H} | \psi \rangle}{\langle \psi | \psi \rangle}$$

always satisfies the inequality

$$E \geq E_0,$$

where $E_0$ is the ground state energy (i.e., lowest eigenvalue) of $\hat{H}$, namely, $\hat{H} |\psi_0\rangle = E_0 |\psi_0\rangle$, with $|\psi_0\rangle$ being the ground state of $\hat{H}$ (i.e., the eigenvector corresponding to the eigenvalue $E_0$).[7]

The basic idea of the method is to build a quantum circuit which can be adjusted to yield as output a state that minimizes $\langle E \rangle$. In other words, one tweaks the gates in the circuit until the output state is close enough to desired ground state $|\psi_0\rangle$. The lower the value of $\langle E \rangle$, the closer one is to $|\psi_0\rangle$.[8]

---

[7]This theorem implicitly assumes that the operator $\hat{H}$ is lower bounded, which is usually expected on physical grounds.

[8]Peruzzo, A., J. McClean, P. Shadbolt, M.-H. Yung, X.-Q. Zhou, P. Love, A. Aspuru-Guzik, J. L. O'Brien. *A variational eigenvalue solver on a photonic quantum processor*. Nat. Comm. 5: 4213

Let $\hat{U} = \hat{U}_m \cdots \hat{U}_2 \hat{U}_1$ be the sequence of $m$ gates making up the variational circuit. Each unitary operator in this sequence represents a gate that is parametrized by a set of "angles" $\{\theta_i\}$, as in $\hat{U}_k = \hat{U}_k(\{\theta_i\})$, with $k = 1, \ldots, m$. For example, consider layers of single-qubit phase gates surrounded by fixed layers of entangling two-quibt gates, the so-called hardware-friendly ansatz, as shown in Fig 12.7.

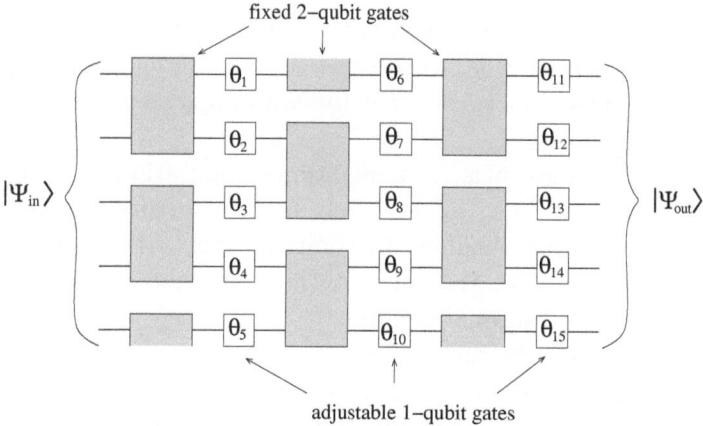

Figure 12.7 Schematic representation of a circuit for variational quantum computing.

Basically, one prepares an initial state $|\Psi_{in}\rangle$, runs it through the quantum circuit $\hat{U}(\{\theta_i\})$, obtain $|\Psi_{out}\rangle = \hat{U}|\Psi_{in}\rangle$, and then measures the expectation value of a Hamiltonian operator $\hat{H}$, namely, $E = \langle\Psi_{out}|\hat{H}|\Psi_{out}\rangle$. One then repeats these steps after making small changes to the parameters, $\theta_i \to \theta_i' = \theta_i + \delta\theta_i$, keeping the same input state $|\Psi_{in}\rangle$ and obtaining as output state $|\Psi_{out}'\rangle$. After computing $E' = \langle\Psi_{out}'|\hat{H}|\Psi_{out}'\rangle$, if $E' < E$, one accepts $\{\theta_i'\}$ as the new parameter values and tries another change starting from these values; if $E' > E$, one discards the changes and try again with a different set of changes, starting again from $\{\theta_i\}$.

There are different strategies to generate the small changes. The generation of changes and analysis of the results (the so-called optimization procedure) can be performed by a classical computer using methods such as gradient descent. As a result, the quantum hardware is just an evaluator of expectation values, which is typically the toughest task to replicate efficiently in a classical computer since it requires dealing with entangled states (i.e., large computational basis).

This method is often called variational quantum eigensolver (VQE) and was first proposed in 2017. There are also several variations of it, some actually dating a few years earlier. They are all examples of *hybrid* approaches, where both classical and quantum hardware are used in combination.

Variational methods have been implemented experimentally in a variety of quantum hardware system and there is already a very large literature about it. These days, when someone talks about doing a quantum computation in hardware, they usually mean implementing a variational quantum algorithm.

Noise also plagues variational quantum computations. Moreover, current hardware only allows for relatively shallow circuits. The large number of measurements required to compute expectation values is also problematic. Critical to the accuracy and efficiency of the method is the choice of the variational circuit.

In addition to materials science and chemistry, VQE and related methods have applications in finances and machine learning (there is some similarity between optimizing the parameters in a variational quantum circuit and training an artificial neural network). So far, results have been interesting but not yet groundbreaking. There are some fundamental challenges to the method (e.g., the insensitivity of the cost function to parameter variations once the circuits become very large, the so-called barren plateau) that researchers around the world are trying to sort out. But we are still in the early days of this approach.

## 12.5   REFERENCES AND FURTHER READING

1. Nielsen M. A. and I. L. Chuang. 2000. *Quantum Computation and Quantum Information.* Cambridge Univ. Press. Sections 7.1, 7.2, and 7.6.

2. Williams, C. P. 2011. *Explorations in Quantum Computing.* Springer-Verlag. Sections 15.1, and 15.4-15.7.

3. Albash, T., and D. Lidar. 2018. *Adiabatic quantum computing.* Rev. Mod. Phys. 90:015002.

4. Lahtien, V. T. and J. K. Pachoes. 2017. *A short introduction to topological quantum computation.* SciPost 3:021.

5. Cerezo, M., A. Arrasmith, R. Babbush, S. C. Benjamin, S. Endo, K. Fuji, J. R. McClean, K. Mitarai, X. Yuan, L. Cincio, P. L. Coles. 2021. *Variational quantum algorithms.* Nat. Rev. Phys. 3:625-644.

# CHAPTER 13

# Quantum Communication

We have seen that Shor's order finding algorithm is capable of factoring semi-prime numbers in polynomial run time. This created a real challenge to some of the most important and widespread public-private key cryptosystems in current use, like RSA and ECC (elliptic curve).

But as much as quantum mechanics breaks cryptography, it can also fix it. There are secure communication protocols based on quantum information processing that allow one to build theoretically unbreakable cryptosystems – at least when all operations are carried out in ideal conditions. The security is not based on the difficulty of solving certain mathematical problems but rather on the physical principles of quantum mechanics.

Once a quantum-based, secure protocol is established for the exchange of information, a network can be established.

To understand how quantum communication protocols work, we will start by introducing a fundamental concept from cryptography. We will then go through the most important protocols in detail. We will then wrap up the chapter with a description of a quantum network for transferring quantum states between quantum sensors and quantum processors.

## 13.1   THE ONE-TIME-PAD CRYPTOSYSTEM

Suppose that Alice wants to send a message $x$ to Bob, where $x \in \{0,1\}^n$. If Eve is eavesdropping on the channel that Alice uses to communicate with Bob, she will be able to retrieve the message, and likely with Alice and Bob never noticing it. If they want to prevent that from happening, they can use a one-time pad encryption. For that, Alice and Bob must share an $n$-bit string of random bit values, $k$, called the key, and known only to them. Then, Alice can encode the plaintext message $x$ into a cipher message $y$ by applying the bit-wise XOR operation

$$y = x \oplus k.$$

Alice sends $y$ to Bob, who can decrypt it back to $x$ by a similar operation, namely,

$$x = y \oplus k.$$

(check it out to make sure you understand it). If Eve intercepts $y$, she has no way of telling what $x$ (the plaintext message) is other than use pure guessing. If $k$ is a truly random bit string, the odds of Eve retrieving the correct (i.e., exact) message are 1 in $2^n$. In other words, it is exponentially hard!

The one-time pad is the most secure cryptosystem possible if:

1. Alice and Bob share a truly random key;

2. the key is used for encryption only once;

3. no one else has access to the key.

While condition (3) requires a high level of trust between the two parties, conditions (1) and (2) can be addressed very objectively. Condition (1) can be satisfied if Alice and Bob use a quantum-based random number generator. Condition (2) means that Alice and Bob must have in storage as many keys as they will ever need to send encrypted messages to each other. But what if they miscalculated the number of keys and now need fresh ones?

This brings to the forefront the so-called key distribution problem: how to securely exchange encryption keys? This is where quantum information processing offers a solution.

## 13.2 QUANTUM KEY DISTRIBUTION

There are several ways to accomplish secure quantum key distribution (QKD). In fact, the main principle predates Shor's algorithm by a decade. We will only survey the most fundamental protocols, starting with BB84.

It is worth noting that QKD is already a technically viable strategy, with some remarkable demonstrations in recent years.

## 13.2.1 BB84

This protocol is named after its inventors, Charles Bennett and Gilles Brassard, and the year of its publication.[1] Here is the sequence of steps:

1. Alice generates a string of random bits, $x \in \{0, 1\}^n$.

2. Alice generates another string of random bits of the same size, $y \in \{0, 1\}^n$.

3. Alice encodes each bit in $x$ as a qubit in the following way:

   - if $y_i = 0$, then

   $$|x_i\rangle = \begin{cases} |0\rangle, & \text{if } x_i = 0 \\ |1\rangle, & \text{if } x_i = 1 \end{cases}$$

   - if $y_i = 1$, then

   $$|x_i\rangle = \begin{cases} \hat{H}|0\rangle, & \text{if } x_i = 0 \\ \hat{H}|1\rangle, & \text{if } x_i = 1 \end{cases}$$

   where $i = 1, \ldots, n$. The $\{|0\rangle, |1\rangle\}$ set is called the $R$ basis and the $\{\hat{H}|0\rangle, \hat{H}|1\rangle\}$ set is called the $D$ basis.

4. Alice sends the $|x\rangle = |x_1\rangle \otimes \cdots \otimes |x_n\rangle$ qubit product state to Bob via a quantum channel (she can send one qubit at a time).

5. Before receiving Alice's qubits, Bob generates his own random string $y' \in \{0, 1\}^n$. He then uses his random string to decide which basis to decode each one of Alice's qubits:

   $$\begin{cases} \text{if} & y_i' = 0 & \Rightarrow & \text{basis } R \\ \text{if} & y_i' = 1 & \Rightarrow & \text{basis } D \end{cases} .$$

6. Alice and Bob share with each other (via a classical public channel) which basis they used for the encoding and measurement, namely, the strings $y$ and $y'$. They discard any bit $x_i$ for which $y_i$ and $y_i'$ do not match. The result is a set of bit values known only to them. That is their shared (but private) key.

---

[1]Bennett, C. and G. Brassard. 1984. *Quantum cryptography: Public key distribution and coin tossing*. Proceedings of the International Conference on Computers, Systems Signal Processing 1: 175–179.

Since Alice and Bob publicly tell each other about their strings $y$ and $y'$, and use an insecure quantum channel for transmitting and receiving the qubits $|x\rangle$, is it possible for some to eavesdrop and obtain the key? The answer is yes for eavesdropping, but no for obtaining the key. Why?

Suppose that Eve intercepts the qubits that Alice is sending to Bob, measures them, and then relays each qubit to Bob according to what she measured. In this process, she may have modified the qubit states or not. Here are the two scenarios:

- if Eve uses the $R$ basis to measure a qubit encoded by Alice in the $R$ basis, the measurement will not modify the qubit, and similarly if the Eve's basis is $D$ and Alice's is $D$;

- if Eve uses the $D$ basis to measure a qubit encoded by Alice in the $R$ basis, the measurement will modify the qubit; for instance, a $|0\rangle$ may become $\frac{1}{\sqrt{2}}(|0\rangle + |1\rangle)$ after Eve's measurement.

- The qubit will also be modified if Eve uses the $R$ basis when Alice encoded the qubit with the $D$ basis.

So, overall, Eve has a 50% chance of using the correct basis. It is a toss up.

When Alice and Bob compare bases, if they find that their bases for a certain qubit are different, they will discard that qubit; if that was a qubit intercepted by Eve, she will have gained no knowledge about the key. But what if Alice and Bob have the same basis? Because of Eve's eavesdropping, there is now a 50% chance that Eve knows the bit value that is going to be incorporated into the key (namely, she may have picked the correct basis). To avoid such a scenario, Bob uses a public channel to send some key bit values back to Alice, who can check if Bob got the correct values. If there is a mismatch, Alice warns Bob and they throw out that entire key, knowing that someone had intercepted it. They then try to obtain a key again. If the number of bits in the string is large, Alice and Bob will eventually get perfect matching of their sacrificial key bits. Eve's chance of success in this scenario becomes exponentially small.

Let us provide a concrete example.

Alice's $x = \{0, 1, 1, 0, 0, 1, 1, 1, 0, 1, 0, 1\}$
Alice's $y = \{D, R, D, R, D, R, D, R, D, D, R, R\}$
Bob's $y' = \{D, R, R, R, D, D, D, D, R, D, R, D\}$

Upon measuring the qubits sent by Alice, Bob finds the values

$$\{0, 1, 0, 0, 0, 1, 1, 0, 1, 1, 0, 0\}.$$

After comparing basis choices, Alice and Bob find out that they had the same basis for qubits 1, 2, 4, 5, 7, 10, and 11. Thus they discard all other bit values, arriving at the following shared key:

$$\{0, 1, -, 0, 0, -, 1-, -, 1, 0, -\}$$

(the discard bit values are replaced by a dash). Now Bob reveals the first three remaining bit values to Alice: 0, 1, 0.

- If Eve eavesdropped and chose the same basis as Alice and Bob $\{D, R, R\}$ for the corresponding qubits, there is no way for Alice and Bob to find out. Eve will know these three bits of the key. Notice that Eve will certainly know that she succeeded because $y$ and $y'$ are publicly known, as well as those three sacrificial bit values that Bob sent to Alice for checking. There is a $2^{-3} = 1/8$ probability of that happening.

- If Eve chose any other basis combination, there is at least a 50% chance that one of Bob's measured qubits will have the wrong value, which Alice will immediately detect. For instance, this would be the case if Eve chose the basis sequence $\{D, R, D\}$ for those three qubits and Bob measured $\{0, 1, 1\}$.

Alice and Bob can keep trying until they are satisfied that no eavesdropping has happened. If they use a large number of bits in the string and a good portion of them for verification, they can reduce Eve's chances of success exponentially! The best Eve can do is to disturb the communication between Alice and Bob by constantly eavesdropping.

The BB84 protocol is very suitable for an optical implementation. The requirements are:

1. a source of single photons (for Alice);

2. a photon polarizer (for Alice);

3. a birefringent crystal and a single-photon detector (for Bob).

All these resources are readily available nowadays. Notice that coherence requirements are minimal (photons can retain their polarization over very long distances when traveling in free space). In addition, no entanglement is necessary.

There is an extension to this protocol where an extra basis is used, namely, $\left\{\frac{1}{\sqrt{2}}(|0\rangle + i|1\rangle), \frac{1}{\sqrt{2}}(|0\rangle - i|1\rangle)\right\}$. It is called the six-state protocol (known as B98) and was proposed by Dagmar Bruss in 1998.[2] It further reduces the probability of success of an eavesdropper.

### 13.2.2 B92

In 1992, Charles Bennett showed that only two orthogonal states are needed for QKD (instead of the four used in BB84 and the six in B98).[3] Here is how this protocol works (we will assume photon qubits).

1. Alice uses a random number generator to create a string of bit values $x \in \{1, 0\}^n$.

2. Alice and Bob previously agreed that if she wants to transmit a 0 ($x_i = 0$), she encodes the qubit as $|0\rangle$; to transmit a 1 ($x_i = 1$), she uses $\hat{H}|1\rangle$ instead. Notice that these states are not orthogonal.

3. Bob generates his own random bit string, $y \in \{0, 1\}^n$.

4. Bob measures the polarization of each qubit sent by Alice with a linear polarizer.[4]

5. If $y_i = 0$, Bob tests for a 0 by setting his polarizer to the orientation $\frac{1}{\sqrt{2}}(|0\rangle + |1\rangle)$. Thus, if Alice sent a 1, the photon will be fully absorbed; if she sent a 0, it will pass through the polarizer with probability $1/2$.

6. If $y_i = 1$, Bob tests for 1 by setting his polarizer to the orientation $|1\rangle$, in which case a 0 will be totally absorbed but a 1 will pass through with probability $1/2$.

---

[2]Bruss, D. 1998. *Optical Eavesdropping in quantum cryptography with six states.* Phys. Rev. Lett. 81: 3018-3021

[3]Bennett, C. H. 1992. *Quantum cryptography using any two nonorthogonal states.* Phys. Rev. Lett. 68: 3121-3124

[4]A linear polarizer lets through a photon without any attenuation only when the photon's polarization is aligned with the main axis of the polarizer; when the main axis of the polarizer is orthogonal to the photon's polarization, the photon is completely absorbed. For any other relative orientation of the polarization with respect to the polarizer's main axis, the photon passes through the polarizers with a probability $0 < p < 1$.

7. Bob sees only 1/4 of the photons pass through. But for the ones that pass through, he can be absolutely sure about their state ($|0\rangle$ and $\hat{H}|1\rangle$, depending on which polarizer he used). Therefore, he just needs to communicate to Alice which qubits passed through, without telling her their values. Alice will know what they are; they will form the shared (private) key.

How to detect eavesdropping in this scheme? Similarly to BB84. Alice and Bob disclose to each other a subset of the key bit values. If the values are identical, they keep the key; if the values are not identical, they can tell with certainty that someone eavesdropped and therefore the key must be abandoned and the protocol restarted.

### 13.2.3  E91

So far, no QKD protocol required entanglement. The first protocol to make use of entanglement was proposed by Artur Ekert in 1991 and works as follows:[5]

1. Alice and Bob share a string of entangled qubit pairs; for instance, $n$ Bell-state pairs $|\Psi\rangle = |B_{11}\rangle$.

2. Alice generates a random $n$-bit string $x \in \{0,1\}^n$.

3. Bob independently generates his own random $n$-bit string $y \in \{0,1\}^n$.

4. Alice measures her qubits as follows: she measures $\hat{X}$ when $x_i = 0$ and $\hat{Z}$ when $x_i = 1$.

5. Bob measures his qubits using $\hat{X}$ when $y_i = 0$ and $\hat{Z}$ when $y_i = 1$.

6. When they happen to choose the same operator for the $i$-th pair, notice that

$$\langle\Psi|\hat{X}_A\hat{X}_B|\Psi\rangle = -1$$
$$\langle\Psi|\hat{Z}_A\hat{Z}_B|\Psi\rangle = -1,$$

which implies that when Alice measures a $+1$, Bob measures a $-1$, and vice versa. When they choose different operators,

$$\langle\Psi|\hat{X}_A\hat{Z}_B|\Psi\rangle = 0$$
$$\langle\Psi|\hat{Z}_A\hat{X}_B|\Psi\rangle = 0,$$

---

[5]Ekert, A. K. 1991. *Quantum cryptography based on Bell's theorem.* Phys. Rev. Lett. 67: 661-663

which means that Alice's and Bob's measurements are uncorrelated.

7. Bob discloses to Alice his operator choices without telling her the actual values of the measurements; Alice tells Bob which choices match hers; they discard all bits where the bases did not match.

8. Bob's sequence of measurements for the remaining qubits is complementary to Alice's, so he just flips the sign of his measurements for those qubits. They now share a private key.

Eavesdropping can be detected similarly to BB84 and B92.

## 13.3 QKD IN PRACTICE

QKD protocols such as BB84 and E91 have been modified for optimal implementation in realistic experimental setups. Other QKD protocols based on the principles of BB84 and E91 have also been invented for the same reason, including protocols that allow one to retrieve more than one bit per photon. There have been multiple experimental implementations of QKD using standard telecommunication optical fibers, with distances starting in the tens of kilometers a couple decades ago and now reaching the thousands through dedicated networks. QKD has also been demonstrated in free space, starting with distances of a few kilometers and now reaching low-orbit satellites.[6] Space agencies of various nations have started QKD satellite programs. Key rates of 100 Mbits per second over 10-kilometer optical fibers are now possible, making QKD a realistic option for certain specialized earthly uses.[7] There exist multi-node QKD-networks in various countries around the globe.

Despite these remarkable advances, QKD still faces several challenges and limitations.

The BB84 protocol relies on Alice and Bob being able to generate perfectly random bit strings; any correlation in the bit string could be exploited to reduce security. Deterministic algorithms for random number generators, like those employed with digital computers, are never truly random (and for this reason are called pseudorandom). Thus, for

---

[6]Lu, C.-Y., Y. Cao, C.-Z. Peng, and J.-W. Pang. *Micius quantum experiments in space.* 2022. Rev. Mod. Phys. 94:035001

[7]Li W., et al. *High-rate quantum key distribution exceeding 110 Mb/s.* Nature Photon. 17:416-421

QKD applications one often relies on quantum hardware-based sources of random bits.

Two other possible issues are known for QKD protocols. One is the so-called person-in-the-middle attack, when Eve impersonates Bob. There is no remedy for that in the BB84 protocol other than Alice and Bob start with a short common secret key in order to recognize each other before implementing the protocol (this is known as source authentication). Alternatively, they can use an asymmetric, public-private cryptosystem for authentication. The other issue is side-channels, which can happen if Alice inadvertently encodes more than one degree of freedom which Eve could measure without Alice's or Bob's knowledge. Fortunately, a solution has been proposed, the so-called measurement-device-independent QKD.[8]

An issue that appears in the implementation of all QKD protocols is noise, which can corrupt the qubits and introduce errors that may be perceived as the result of eavesdropping. Two techniques have been devised to deal with such a situation.

- Error reconciliation: Alice and Bob need to fix key discrepancies without revealing information to Eve. They can use parity checking protocols such as CASCADE.[9]

- Privacy amplification: to make any discovery more difficult for Eve, Alice and Bob can use their error-reconciled key to generate another, shorter key. For that, they can use pre-established hash functions. (Hash functions map inputs of any size to fixed-size output values. Cryptographically-secure hash functions exist and they can be used for authentication.)

In practice, it is believed that the best approach is to combine the one-time-pad afforded by QKD (which consumes lots of qubits but is presumed secure) with a classical block cipher such as AES (which is very economical but not known to be absolutely secure). The one-time pad works by rekeying AES every so often, making it more difficult for an attacker, as the key is constantly changing.

Finally, even though, ideally, the security of QKD is guaranteed by

---

[8]Lo, H.-K., M. Curty, and B. Qi. *Measurement-device-independent quantum key distribution.* 2012. Phys. Rev. Lett. 108:130503

[9]Martinez-Mateo, J., C. Pacher, M. Peev, A. Ciurana, and V. Martin. 2014. *Demystifying the information reconciliation protocol CASCADE.* arXiv:1407.3257

the laws of physics, in reality it is not unconditional. It is highly dependent on the accuracy of the hardware design and its implementation. At the moment, the error rates and vulnerabilities induced by hardware are still too high to satisfy the stringent security criteria needed for the most sensitive cryptography applications, such as national security.[10]

## 13.4   QUANTUM NETWORKS

QKD offers a solution to a very important practical problem by exploiting quantum mechanical principles such as superposition, uncertainty, and possibly entanglement. But once the secret key has been established and shared securely, all encrypted communications can proceed classically, namely, without resorting to interference or entanglement and using standard classical channels.

What if we need to transmit quantum states on a regular basis? For instance, when interconnecting quantum sensors or distributing a quantum algorithm workload over several quantum processors? For these purposes, the concept of quantum networks was introduced. The basic idea is to gather, process, and store quantum information locally in quantum nodes, and use photons as flying qubits to transport quantum states from one node to another, similarly to what is done in the Internet. The photons can be in the optical range (e.g., traveling through optical fibers) or in the radio and microwave range (traveling through free space).[11] Just like in the case of the Internet, in addition to transmission channels, a quantum network requires components such as transmitters, receivers, memories, and repeaters, as well as transducers. However, differently from the Internet, these components must operate in phase-coherent conditions, allowing entanglement to be distributed and preserved. While single-photon emitters and detectors are readily available, quantum memories and repeaters are still being researched, tested, and perfected. Transducers are needed to convert quantum signals of a frequency range into another (e.g., microwave to optical) and are also being intensively researched and developed.

It is difficult to discuss quantum networks without going into technical details of how the physical realizations of various components work.

---

[10]https://www.nsa.gov/Cybersecurity/Quantum-Key-Distribution-QKD-and-Quantum-Cryptography-QC/

[11]Fiber optic wavelengths for telecommunications are usually 850, 1300, and 1500 nm, falling in the infrared; microwave frequencies range from 300 MHz to 300 GHz, with corresponding wavelengths going from 1 m to 1 mm.

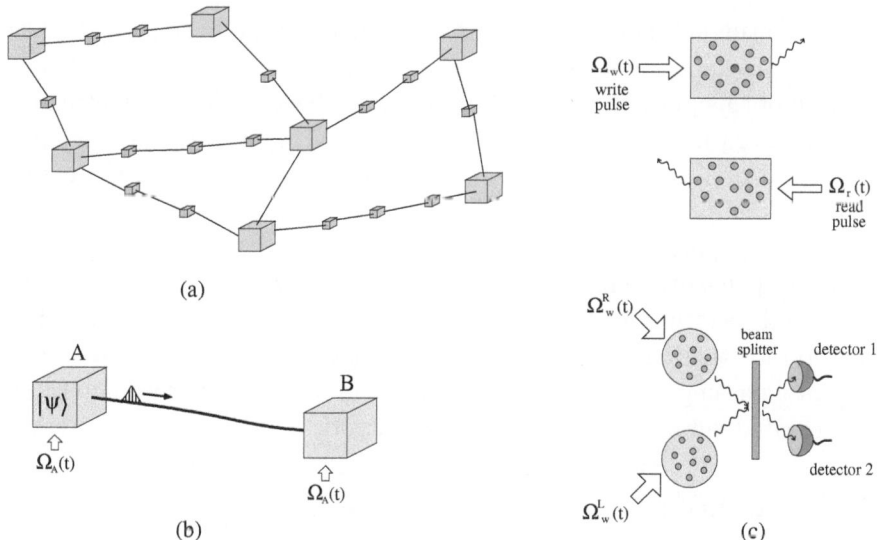

(a)

(b)

(c)

Figure 13.1  (a) Illustration of a quantum network where quantum states are transmitted from one node to another. The larger cubes represent quantum nodes and the solid lines are quantum channels. The smaller cubes are quantum repeaters. (b) A protocol for transferring a quantum state $|\psi\rangle$ from node A to node B via a flying photon qubit. (c) Schematic illustration of the DLCZ protocol for a quantum repeater: write and read operations on an atomic gas ensemble (which functions as a quantum memory), with the emission of a herald photon (upper panel); their use for entangling two ensembles (lower panel).

Therefore, here we will confine ourselves to some general concepts. References at the end of the chapter provide in-depth, technical descriptions. The layout of a quantum network is illustrared in Fig. 13.1a while Fig. 13.1b shows a protocol for transferring a quantum state $|\psi\rangle$ from node A to node B.[12] Suitably timed pulses $\Omega_A(t)$ and $\Omega_B(t)$ are employed to induce transitions within nodes A and B (typically atomic cavities) which emit and absorb photons, respectively. These photons travel through the quantum channel connecting the two nodes. The diagrams in Fig. 13.1c illustrate a protocol (named DLCZ, after its inventors) for

[12]Cirac, J. I., P. Zoller, H. J. Kimble, and H. Mabuchi. 2007. *Quantum state transfer and entanglement distribution among distant nodes in a quantum network.* Phys. Rev. Lett. 78:3221-3224

implementing a quantum repeater.[13] A low-intensity "write" light pulse hits an ensemble of $N$ identical atoms kept at their ground states. A photon of this pulse is absorbed by the ensemble creating a quantum superposition where $N - 1$ atoms remain in the ground state and one atom is excited to a long-living state. During the excitation process, a herald photon is emitted by the ensemble, signaling that the system is storing the desired quantum state. To retrieve the stored state, another weak pulse is sent to the ensemble, taking all $N$ atoms back to their ground state and resulting in the emission of another herald photon. Using this method for writing and reading states in parallel for two atomic ensembles and passing the emitted photon pair through an interferometer, one can created entangled states across a network.

## 13.5 REFERENCES AND FURTHER READING

1. Williams, C. P. 2011. *Explorations in Quantum Computing.* Springer-Verlag. Chapter 13.

2. Nielsen M. A. and I. L. Chuang. 2000. *Quantum Computation and Quantum Information.* Cambridge Univ. Press. Section 12.6.

3. Mermin, N. D. 2007. *Quantum Computer Science.* Cambridge Univ. Press. Section 6.2.

4. Diamanti, E., H.-K. Lo, B. Qi, and Z. Yuan. 2016. *Practical challenges in quantum key distribution.* npj Quantum Inf. 2:16205

5. Kimble, H. J. 2008. *The quantum internet.* Nature (London) 453:1023-1030

6. Gisin, N. and R. Thew. 2007. *Quantum Communication.* Nature Photon. 1:165-171

7. Lauk, N., N. Sinclair, S. Barzanjeh, J. O. Covey, M. Saffman, M. Spiropulu, and C. Simon. 2020. *Perspectives on quantum transduction.* Quantum Sci. Technol. 5:020501

---

[13]Duan, L.-M., M. D. Lukin, J. I. Cirac, and P. Zoller. 2001. *Long-distance quantum communication with atomic ensembles and linear optics.* Nature (London) 414:413-418

## 13.6   EXERCISES AND PROBLEMS

1. Reformulate the E91 protocol to employ the $|B_{00}\rangle$ Bell state.

2. (Adapted from C. P. Williams, Explorations in Quantum Computing, 2nd edition.) The BB84 protocol employs single, non-entangled qubits to perform QKD, while the E91 protocol is based on entangled qubit pairs. Despite this physical difference, they are surprisingly similar.

   (a) Verify that the following identity holds for arbitrary one-qubit gates $\hat{U}_1$ and $\hat{U}_2$:

   $$(\hat{U}_1 \otimes \hat{U}_2)|B_{00}\rangle = \hat{I} \otimes (\hat{U}_2 \hat{U}_1^{\ T})|B_{00}\rangle,$$

   where $|B_{00}\rangle$ is the first Bell state and the superscript $T$ denotes transposition. *Hint*: represent operators and vectors in the two-qubit computational basis.

   (b) Explain how this identity can be used to find a connection between BB84 and E91.

# CHAPTER 14

# Quantum Sensing

In quantum sensing, one uses the properties of a quantum system such as energy quantization, interference, or entanglement to detect and measure physical quantities of other systems. The goal is to achieve higher sensitivity, precision, and resolution than what is possible using standard (non quantum) devices. For that, one exploits the high sensitivity quantum states have to external perturbations.

One can say that quantum sensing is the oldest form of information processing reliant on quantum properties. For decades, superconducting quantum interference devices (SQUIDs) have been used to measure weak magnetic fields; atomic clocks were introduced even before SQUIDS. But the area has greatly progressed since those pioneering devices and a variety of new quantum sensors, often at a single-atom level, have entered the scene in recent times, vastly improving and expanding the applications of quantum sensing. New terminologies have also been introduced, such as "quantum metrology" and "quantum-enhanced sensing" to more clearly indicate the key role played by unique quantum properties such as entanglement. While quantum computing gets most of the attention these days, in reality quantum sensors will have a much more profound economical impact in the short and long terms given their wide range of uses.

## 14.1 TYPES OF QUANTUM SENSING

There are three types of quantum sensing, depending on the particular quantum property utilized by the sensor:

1. energy quantization sensors (i.e., sensing relying on the existence of discrete energy levels),

2. quantum coherence sensors (i.e., sensor exploiting the ability of a quantum system to retain spatial or temporal superpositions), and

3. quantum entanglement sensors (multi-component sensors that use the sensitivity of entanglement to external perturbation).

Whatever physical system is chosen to realize a quantum sensor, it must satisfy certain criteria to properly function:[1] energy levels must be well defined and sufficiently separated to be distinguishable; the system must be initializable in a suitable quantum state and its states accessible to readout; the system must remain coherent during manipulations (i.e., decoherence times must be longer than the operation time); and the interaction of the system with a suitable external field must cause detectable internal changes. The latter two attributes are the hardest to achieve: the system must be very sensitive to a particular signal and at the same time be immune to noise. Often, these conditions compete against each other.

## 14.2 A QUANTUM SENSING PROTOCOL

Because of the largest variety of quantum sensors and their uses, protocols for their operation can be bespoke. However, there is a generic theoretical framework which fits most cases of type (1) or (2) and which helps one understand the principles involved in quantum sensing. It starts by setting up the Hamiltonian of the sensor:

$$\hat{H}(t) = \hat{H}_0 + \hat{H}_V(t) + \hat{H}_{\text{ctrl}}(t),$$

where $\hat{H}_0$ is the internal Hamiltonian, $\hat{H}_V(t)$ encodes the interaction with an external signal $V(t)$, which is the quantity to be measured, and $\hat{H}_{\text{ctrl}}(t)$ represents the control exerted on the sensor by the operator. The signal $V(t)$ affects the sensor's dynamics through $\hat{H}_V(t)$, which is to be counterbalanced by $\hat{H}_{\text{ctrl}}(t)$. Out of this counterbalance, one infers $V(t)$.

In its simplest form, a quantum sensor may consist of a two-level system (a qubit):

$$\hat{H}_0 = E_0 |0\rangle\langle 0| + E_1 |1\rangle\langle 1|,$$

where $E_0$ and $E_1$ are the energy eigenvalues associated to the eigenstates $|0\rangle$ and $|1\rangle$, respectively. Here, we will adopt these two eigenvectors as a

---

[1]This also applies to qubits used in quantum computing and quantum communications; see Sec. 15.1.

computational basis and assume $\hbar\omega_0 = E_1 - E_0 > 0$. Then, an appropriate signal Hamiltonian has the form

$$\hat{H}_V(t) = \gamma \left[ V_x(t)\,\hat{X} + V_y(t)\,\hat{Y} + V_z(t)\,\hat{Z} \right],$$

where $\gamma$ is the so-called transduction parameter and $V_x$, $V_y$, and $V_z$ are the different components of the external field with respect to the space of Pauli operators specific to the qubit. $\hat{H}_{\text{ctrl}}(t)$ encodes the application of standard quantum gates such as Hadamard or a sequence of pulses, such as $\pi/$ and $\pi$.

Once a system is established in these terms, the protocol for obtaining $V(t)$ consists of the following steps:

1. Initialize the sensor, say, at state $|0\rangle$.

2. Use pulses to bring the sensor to a desired sensing state: $|\psi_0\rangle = \hat{U}_{\text{prep}}|0\rangle$.

3. Let the sensor evolve for a time $t$, when $\hat{H}_0$ and $\hat{H}_V(t)$ are acting on it, to reach the state

$$|\psi(t)\rangle = \hat{U}_H |\psi_0\rangle = c_0|\psi_0\rangle + c_1|\psi_1\rangle,$$

where $\hat{U}_H$ the evolution operator for that time period and $|\psi_1\rangle$ is a state orthogonal to $|\psi_0\rangle$. The evolution induces a superposition of the $|\psi_0\rangle$ and $|\psi_1\rangle$ states.[2]

4. Apply the inverse sequence of pulses to return the sensor back to the computational basis:

$$|\psi_{\text{final}}\rangle = \hat{U}^\dagger_{\text{prep}}|\psi(t)\rangle = c_0|0\rangle + e^{i\theta} c_1|1\rangle,$$

where $\theta$ is an unimportant phase.[3]

5. Read out the final state of the sensor.

Upon repeating steps 1-5 multiple times, one obtains an estimate of the transition probability $p$,

$$p = 1 - |c_0|^2 = |c_1|^2.$$

---

[2] Notice that because of the two-qubit state, given $|\psi_0\rangle$, $|\psi_1\rangle$ is unique up to a phase factor.

[3] As an exercise, the reader is invited to derive this relation.

To infer $V(t)$ from $p$, one can repeat the protocol varying the time interval $t$ or any other parameter that may help deconvolute $V(t)$ from a set of estimated probabilities.

To illustrate the procedure, consider the following example known as Ramsey measurement, where the goal is to measure a static longitudinal perturbation $V(t) = V_z \hat{Z}$:

1. Initialize the sensor in the state $|0\rangle$.

2. Apply a $\pi/2$ pulse to bring the sensor to the superposition state

$$|\psi_0\rangle = \frac{1}{\sqrt{2}} (|0\rangle + |1\rangle).$$

3. Let the sensor evolve for a time $t$, such that its state becomes

$$|\psi(t)\rangle = \frac{1}{\sqrt{2}} \left( |0\rangle + e^{-i\omega t} |1\rangle \right),$$

where global phases have been neglected and $\omega = \omega_0 + \gamma V_z/\hbar$.

4. Apply another $\pi/2$ pulse to bring the sensor back to the original state, resulting in

$$|\psi_{\text{final}}\rangle = \frac{1}{2} \left[ \left( 1 + e^{-i\omega t} \right) |0\rangle + \left( 1 - e^{-i\omega t} \right) |1\rangle \right].$$

5. Read out the final state in the computational basis.

Upon repeating the procedure and sweeping over a range of time periods, one can obtain the transition probability $p$ as a function of time, which should follow the relation

$$p(t) = \frac{1}{2} - \frac{1}{2} \cos(\omega t).$$

One can then extract $V_z$ by subtracting $\omega_0$ from $\omega$. Ideally, one could start with a qubit system where $\omega_0 = 0$, in which case the oscillation frequency corresponds directly to $V_z/\hbar$. These oscillations are known as Ramsey fringes.

A similar procedure can be employed to measure a static transverse field $V(t) = V_x \hat{X}$. In this case, there is no need for steps #2 and #4 and the oscillations of the time-dependent transition probability, known as Rabi oscillations, directly determine $V_x$.

## 14.3 ENTANGLEMENT-BASED QUANTUM SENSORS

While single-qubit sensors can beat the sensitivity of classical devices, even better results can be obtained when an ensemble of those sensors is employed. In the context of metrology, one calls the standard quantum limit the maximum precision achievable when the qubits in the ensemble are uncorrelated. In this limit, for a fixed measurement time, the precision scales as $1/\sqrt{N}$, where $N$ is the number of qubits in the ensemble. The gain by employing more than one qubit in this case is merely statistical. However, upon allowing the qubits to interact and build a fully entangled state, one can go beyond that and move toward the so-called Heisenberg limit, where the precision scales as $1/N$. There has been recent progress in exploring this effect on ensembles neutral atoms. There are also proposals for employing quantum networks to entangle atomic clocks and enhance their performance, with applications ranging from better GPS systems to scientific, such as the detection of gravitational waves and dark matter.

One can understand the advantage of employing entangled states by going back to the protocol of Sec. 14.2 and using instead of a single qubit, an ensemble of $N$ qubits initially prepared in the $|0 \cdots 0\rangle$ state and then brought to the sensing state (known as the GHZ state)

$$|\psi_0\rangle = \frac{1}{\sqrt{2}} \left( |0 \cdots 0\rangle + |1 \cdots 1\rangle \right).$$

Using the same $\hat{H}_0$ Hamiltonian for each qubit, the $N$-qubit state picks up a phase enhanced by a factor of $N$, namely, $N\omega_0 t$ after evolving for a time $t$. The end result of the protocol is that the transition probability of any of the qubits (and we need to measure only one of them) becomes $p = \sin^2(N\omega_0 t/2)$, indicating an increase in the oscillation frequency by a factor of $N$, and, consequently, a reduction in the measurement time by the same factor.

A common approach for implementing entanglement-based sensing is to use spin squeezed states. These are states prepared in such a way that the uncertainty associated to one angular momentum component is reduced at the expense of increasing the uncertainty of the others (hence the connection to "squeezing"). Indeed, applying Eq. (4.1) to the Cartesian components of the total spin angular momentum operator, which satisfy the commutation relation $[\hat{J}_x, \hat{J}_y] = i\hbar\hat{J}_z$, we find

$$\left( \Delta J_x \right) \left( \Delta J_y \right) \geq \frac{\hbar}{2} |\langle \hat{J}_z \rangle|.$$

Thus, if the state is prepared such that $\Delta J_x < \sqrt{\hbar |\langle J_z \rangle|}/2$, then $\Delta J_y \geq \sqrt{\hbar |\langle J_z \rangle|}/2$. Usually, achieving this situation is only possible through entanglement. In this case, one can take advantage of the smaller uncertainty in $\hat{J}_x$ to improve sensitivity. Going back to the protocol of Sec. 14.2, an appropriate "squeezing" Hamiltonian must then be employed during the state preparation.

The most critical issue for entanglement-based quantum sensing is decoherence and noise. Left unchecked, noise can remove all the scaling advantages of entanglement. This challenge is being tackled in a number ways, from using specially engineered entangled states that are robust to the prevailing noise type to the application of quantum error correction.

## 14.4 REFERENCES AND FURTHER READING

1. Degen, C. L., F. Reinhard, and P. Cappellaro. 2017. *Quantum sensing.* Rev. Mod. Phys. 89:035002

2. Ye, Jun and P. Zoller. 2024. *Essay: Quantum sensing with atomic, molecular, and optical platforms for fundamental physics.* Phys. Rev. Lett. 132:190001

3. Bongs, K., S. Bennett, and A. Lohmann. 2023. *Quantum sensors will start a revolution – if we deploy them right.* Nature 617:672-675

# CHAPTER 15

# Physical Realizations of Qubits

In the mid 1990s, with the advent of quantum algorithms that could solve hard problems of practical interest, a race to develop qubits began. A myriad of approaches were proposed and tried, from nuclear spins in molecules to trapped ions and electrons to superconducting circuits, not to mention photons. Some of these qubit realizations were far more advanced than others from a technical standpoint (such as nuclear spins and superconducting circuits), thanks to decades of fundamental studies and applications in areas other than quantum computing. Other proposals were untested but very promising such as defect states in insulating materials and Rydberg atomic systems.

By the 2010s, it was clear that ion traps and superconducting circuits offered the best tradeoff between scalability and quality, at least in the short term. Much was invested in their development and they are, by now, far more established than other physical realizations of quantum processors. Yet, other approaches continue to be pursued either because of their promise of superior qubit quality or their ability to scale, if successful. An example of the latter are qubits based on defects in crystal lattices. Recently, quantum processors based on Rydberg atoms have also made tremendous progress and are now commercially available. Photonic qubits continue to evolve. The verdict to which physical realization will prevail is still unclear, including the possibility of something entirely new. It is likely that different realizations will find its own niche use.

Below, we provide brief explanations of the inner workings of the prevailing types of qubits today. But before going into that, we list what is expected of a good qubit.

## 15.1   DIVINCENZO'S CRITERIA

In 2000, David DiVincenzo proposed a series of requirements that any serious quantum computing hardware contender needs to satisfy.[1] They are now called DiVincenzo's criteria and are essentially the following:

1. A scalable system must have well-defined qubits (i.e., well-defined two-level quantum states).

2. It must be possible to initialize qubits to a fiducial state, say, $|0\rangle$.

3. Decoherence times must be much larger than the gate operation times.

4. A universal set of quantum gates must be possible (e.g., $H$, $T$, CNOT).

5. It must be possible to readout qubit states, i.e., make measurements.

For quantum communication hardware, two additional criteria are necessary:

1. It must be possible to transmit flying qubits between separate locations.

2. It must be possible to convert stationary to flying qubits, and vice-versa.

We will now look into a few physical systems that satisfy DiVincenzo's criteria.

## 15.2   TRAPPED IONS

The quantized energy levels in atoms are a natural qubit basis, provided that the lowest-energy states can be isolated from other states. However, addressing individual atoms and controlling their interactions is very hard for neutral atoms.[2] The solution is to use charged atoms, namely, ions. Since the 1950s, techniques for trapping individual ions have been

---

[1]DiVincenzo, D. P. 2000. *The physical implementation of quantum computers.* Fortschr. Phys. 48: 9-11

[2]An exception is the so-called Rydberg atoms, which have very large orbital states. See Sec. 15.5.

developed and constantly improved, starting with the Paul trap.[3] It is impossible to confine charged particles in three-dimensional space using solely static electric fields.[4] However, one can build a configuration of time-dependent electric fields that, on average, confine in all three spatial dimensions. If the switching between confining and anti-confining potentials happens sufficiently fast, faster than the time it takes for the particle to escape, the particle remains trapped inside a well defined region. That is the principle of a Paul trap. One way to visualize this effect is to consider a rotating saddle, as shown schematically in Fig. 15.1

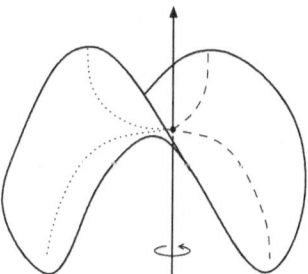

Figure 15.1  Schematic representation of a rotating saddle potential.

The saddle potential can be formed by an alternating quadrupole field configuration, as the one shown in Fig. 15.2

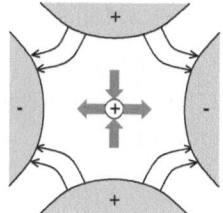 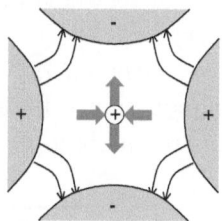

Figure 15.2  Alternating quadrupole field configurations.

Given the particle's mass, electrical charge, and the profile of the saddle potential, there is a range of frequencies such that the particle remains confined in a stable configuration. For ions, this typically happens in the radiofrequency range.

---

[3]After Wolfgang Paul, who invented it in 1953 and received a Nobel Prize for it in 1989.

[4]This is known as Earnshaw's theorem and is a direct consequence of Gauss' law. There can be no stable equilibrium point in a free electromagnetic static field, only saddle points.

In order to be able to individually access ions (each ion being a qubit), they need to be sufficiently apart, so that laser beams can be pointed at each one without affecting its neighbors. Since most transitions of relevance to ion qubits are in the visible range of the electromagnetic spectrum, the lasers employed to manipulate qubit states have wavelengths in the micrometer range, and that is the typical spatial separation of trapped ions, see Fig. 15.3.

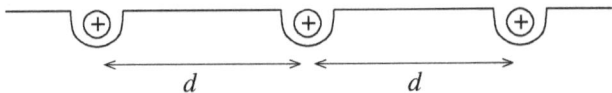

Figure 15.3  Ion trapping in one dimension. In practice, $d \approx 1$ $\mu$m.

But that alone is not enough to operate an ensemble of ions as qubits. They also have to be cooled down to very low temperatures. This requires the emission and absorption of vibrations at the individual ion level, avoiding disturbing other ions. By cooling the ions (i.e., reducing their jiggling motion) and exploring their long-range Coulomb interactions with each other, one can turn the ions in a trap into a "rigid" system, like atoms in a crystal, with very little relative motion. As a result, photons of certain wavelengths can only be absorbed by the totality of the ions; the photon linear momentum is then distributed among all ions, causing only a very small collective displacement (this is called the Mössbauer effect).

Let us consider the energy levels of a $Ca^+$ ion shown in Fig. 15.4, which is a common choice for ion traps. It has a single electron in its outer shell, thus resembling a bit a hydrogen atom.

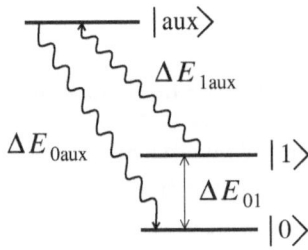

Figure 15.4  Energy levels in a $Ca^+$ qubit. The energy scales are: $\Delta E_{01} = 1.69$ eV, $\Delta E_{1-aux} = 1.45$ eV, and $\Delta E_{0-aux} = 3.14$ eV. The corresponding photon wavelengths are 732 nm, 854 nm, and 393 nm, respectively.

The ground state plays the role of the $|0\rangle$ state. The first excited state is the $|1\rangle$ state. The latter is metastable, with a lifetime of approximately 1 s. This seems short, but in practice is a very long time. A third, auxiliary higher-energy state, $|aux\rangle$, with a very short lifetime (7 ns) is used for measurement purposes, as well as for other operations.

The ground state of a $Ca^+$ ion is actually double degenerate, so optical pumping is needed to select a single low-energy $|0\rangle$ state, as shown in Fig. 15.5.

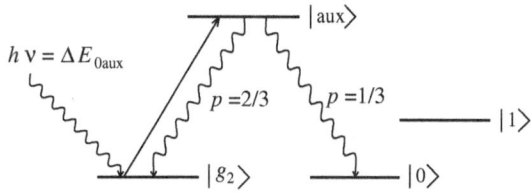

Figure 15.5  Qubit initialization via optical pumping in $Ca^+$ qubits.

By sending only photons with frequency $\Delta E_{0-aux}/c$, with the appropriate polarization, transitions between the unwanted ground state $|g_2\rangle$ and the auxiliary state $|aux\rangle$ are induced, but not between the $|0\rangle$ and $|aux\rangle$ states. The auxiliary state decays back to one of the ground states; when it decays to $|0\rangle$, it stays there, but when it decays to the unwanted state, it is eventually excited again. Therefore, over time, the ions converge toward the $|0\rangle$ state.

The auxiliary state and its fast decay can be explored for performing measurements: by sending a photon with frequency $\Delta E_{1-aux}/c$, if the qubit is in the state $|1\rangle$, the state $|aux\rangle$ is excited and rapidly decays toward the ground states, emitting a photon that can be detected; if no photon is emitted and detected, one can infer that the qubit was in the state $|0\rangle$ and therefore was not excitable by the incoming photon. Alternatively, if a photon of frequency $\Delta E_{0-aux}$ is used, it will stimulate an excitation from $|0\rangle$ to $|aux\rangle$, ensuing a rapid response (i.e., the emission of a photon); in this case, if the qubit was in the state $|1\rangle$, no emission is expected.

What about qubit operations?

Single-qubit operations can be performed by sending laser pulses with frequencies tuned to $\Delta E_{01}$ and with a duration corresponding to a particular Bloch vector angle to perform a rotation in the Bloch sphere. For instance, a $\pi$ pulse can take from 1 to 10 $\mu$s. We will skip the derivation of the effective Hamiltonian behind such a qubit operation when a resonant laser is employed. The resulting evolution operator, in the standard computational basis, is

$$\hat{U}(t,0) = \begin{pmatrix} e^{i\omega_0 t/2} \cos\Omega(t) & ie^{i(\delta+\omega_0 t/2)} \sin\Omega(t) \\ ie^{-i(\delta+\omega_0 t/2)} \sin\Omega(t) & e^{-i\omega_0 t/2} \cos\Omega(t) \end{pmatrix},$$

where $\Omega(t) = \Delta \times t/2$, $\omega_0 = \Delta E_{01}/\hbar$, $\delta$ is a laser phase parameter, and $\Delta$ is a laser-atom coupling parameter. For a pulse of duration $t = 3\pi/\omega_0$, $\Delta = \omega_0/6$, and $\delta = 3\pi/2$, one finds

$$\hat{U} = -\frac{i}{\sqrt{2}} \begin{pmatrix} 1 & 1 \\ 1 & -1 \end{pmatrix} = -i\hat{H},$$

namely, a Hadamard gate up to an overall phase factor. Other gates, including phase gates can be obtained in a similar faction, by a proper choice of parameters.

Two-qubit gate operations are more challenging but can also be performed by exploiting the inter-ion interactions. As mentioned earlier, because the ions are trapped and cooled, the only residual interaction left is their mutual Coulomb repulsion. The ions, when confined in a linear trap, form a chain which can hold quantized vibrational modes, the so-called phonons. The phonon energy is given by

$$E = n\,\hbar\,\omega_{\mathrm{ph}},$$

where $n = 0, 1, 2, \ldots$ and $\omega_{\mathrm{ph}}$ is an angular frequency related to the stiffness of the linear chain and the ion masses. We have to think of such a system as a collection of qubits attached to quantized phonon modes,

$$H = H_q \otimes H_{\mathrm{ph}}$$

(i.e., the product of two Hilbert spaces). The phonons can intermediate interactions among qubits and in a controllable way. How?

Consider what happens when we shine light with angular frequency $\omega_0 \pm \omega_{\mathrm{ph}}$ on an ion:

$$|0\rangle_q \otimes |n\rangle_{\mathrm{ph}} \quad \xrightarrow[\omega_0 + \omega_{\mathrm{ph}}]{} \quad |1\rangle \otimes |n+1\rangle \quad \text{"blue sideband"}$$

$$|0\rangle_q \otimes |n\rangle_{\mathrm{ph}} \quad \xrightarrow[\omega_0 - \omega_{\mathrm{ph}}]{} \quad |1\rangle \otimes |n-1\rangle \quad \text{"red sideband"}.$$

We can alter the individual qubit state at the same time we either increase or decrease the number of phonon modes in the chain. By pulsing the laser at these frequencies in a selective way, we can create entangled superpositions between an ion and a phonon:

$$|0\rangle_q \otimes |0\rangle_{\text{ph}} \longrightarrow \frac{1}{\sqrt{2}}(|0\rangle_q \otimes |0\rangle_{\text{ph}} + |1\rangle_q \otimes |1\rangle_{\text{ph}}).$$

Bringing a second ion, we can have

$$|0\rangle_{q_A} \otimes |0\rangle_{q_B} \otimes |0\rangle_{\text{ph}} \longrightarrow \frac{1}{\sqrt{2}}(|0\rangle_{q_A} \otimes |0\rangle_{q_B} \otimes |0\rangle_{\text{ph}} + |1\rangle_{q_A} \otimes |0\rangle_{q_B} \otimes |1\rangle_{\text{ph}}).$$

Sending a second laser pulse, this time targeting the second ion and a resonance frequency $\omega_0$ (the so-called carrier frequency), we get

$$\frac{1}{\sqrt{2}}(|0\rangle_{q_A} \otimes |0\rangle_{q_B} \otimes |0\rangle_{\text{ph}} + |1\rangle_{q_A} \otimes |0\rangle_{q_B} \otimes |1\rangle_{\text{ph}})$$

$$\longrightarrow \quad \frac{1}{\sqrt{2}}(|0\rangle_{q_A} \otimes |1\rangle_{q_B} \otimes |0\rangle_{\text{ph}} + |1\rangle_{q_A} \otimes |1\rangle_{q_B} \otimes |1\rangle_{\text{ph}}).$$

Now we can send a third laser pulse, also targeting the second ion but at a frequency $\omega_0 - \omega_{\text{ph}}$, leading to

$$\frac{1}{\sqrt{2}}(|0\rangle_{q_A} \otimes |1\rangle_{q_B} \otimes |0\rangle_{\text{ph}} + |1\rangle_{q_A} \otimes |1\rangle_{q_B} \otimes |0\rangle_{\text{ph}})$$

$$= \quad \frac{1}{\sqrt{2}}(|0\rangle_{q_A} \otimes |1\rangle_{q_B} + |1\rangle_{q_A} \otimes |0\rangle_{q_B}) \otimes |0\rangle_{\text{ph}}.$$

The net result is a two-qubit gate between the two ions:

$$|0\rangle_{q_A} \otimes |0\rangle_{q_B} \otimes \longrightarrow \frac{1}{\sqrt{2}}(|0\rangle_{q_A} \otimes |1\rangle_{q_B} + |1\rangle_{q_A} \otimes |0\rangle_{q_B}),$$

which is a maximally entangled state. This approach is known as the Cirac-Zoller mechanism and can be extended to produce other gates and states.[5] Nowadays, in practice, another approach is used, the Mølmer-Sørensen procedure.[6] It employs two detuned lasers (a "blue" and a "red" one), $\omega_{\text{laser}} = \omega_0 \pm \delta\omega$. This combination induces an effective interaction among the ions that resembles that between magnetic dipoles,

$$\hat{H} = J_{ij}^x \hat{X}_i \hat{X}_j + J_{ij}^y \hat{Y}_i \hat{Y}_j + J_{ij}^z \hat{Z}_i \hat{Z}_j,$$

---

[5]Cirac, J. I. and P. Zoller. 1995. *Quantum computations with cold trapped ions.* Phys. Rev. Lett. 74:4091-4094

[6]Sørensen, A. and K. Mølmer. 1999. *Quantum computation with ions in thermal motion.* Phys. Rev. Lett. 82:1971-1974

where the couplings $J_{ij}^\alpha$ are controlled by laser parameters and the detuning $\delta\omega$. (This type of Hamiltonian is called the Heisenberg model.) The biggest advantage of the Mølmer-Sørensen method is that it is less sensitive to the collective motion of the ion chain, thus it can be employed even when the chain is not at rest. Also, it is faster than the Cirac-Zoller method, which is an important quality given that the states, including those with phonons, can lose coherence. That advantage stems from the fact that ions do not need to be individually addressed, therefore saving operation time. It also covers a wider range of unitary two-qubit operations.

There are other ways to create states $|0\rangle$ and $|1\rangle$ in an ion. For instance, via a hyperfine coupling to the nuclear spin. In $Yb^+$, the lowest-lying electronic state is split into two due to the hyperfine coupling, as shown schematically in Fig. 15.6

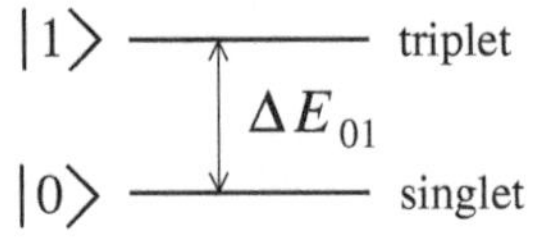

Figure 15.6 The hyperfine split in $Yb^+$, where $\Delta E_{01} \approx 58$ $\mu$eV, with a corresponding photon wavelength equal to 2.3 cm (12.6 GHz).

In this case, the state $|1\rangle$ has an infinite lifetime and transitions between $|0\rangle$ and $|0\rangle$ are induced by microwave pulses, without the need of any auxiliary intermediate state. The state $|1\rangle$ can be further split by an external magnetic field. (In practice, one still uses an auxiliary state to excite $|0\rangle - |1\rangle$ transitions.) The biggest advantage of using hyperfine ion qubits is their resilience to decoherence, which leads to much longer tie spans to implement sequences of quantum gates.

Ion trap qubits have made tremendous progress since their inception in the late 1990s. There are currently ion-based quantum processors with more than 100 qubits. Yet, they do have some intrinsic issues. The main one is the reliance on vibrational modes in the ion chair to implement two-qubit gates. The more ions you pack together, the closer the vibrational frequencies become ($\omega_{ph} \sim 1/L$, where $L$ is the chain length). Small frequencies make tuning hard, and at some point impossible. Trying to compensate by employing weaker coupling slows down the gates and reduces the number of operations within the coherence time. To address this issue, shorter linear chairs coupled at their endpoints have been proposed. A example is the quantum charge-coupled device (QCCD), where certain ions are shuttled from one chain to another, as illustrated in Fig. 15.7.

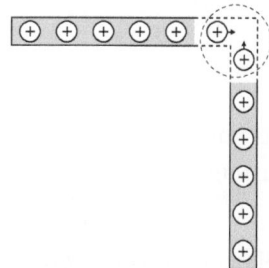

Figure 15.7 Shuttling ions between two chains to perform two-qubit operations. The dotted line encircles the interaction region.

However, shuttling ions creates its own problem, as charge in motion couples to a variety of environmental degrees of freedom, inducing additional decoherence channels. So far, the best QCCD architecture can achieve only six qubits and two interaction zones. Alternative architectures keep the long chains and try to use pulse engineering to shorten the gate duration.

Another issue of ion trap qubit systems is photon loss. One way to remediate it is to employ microwave cavities.

Nevertheless, these difficulties have not prevented the commercialization of ion-trap quantum technologies. Several companies are active in this area.

## 15.3 SUPERCONDUCTING CIRCUITS

All qubits based on superconducting circuits exploit a phenomenon called Josephson effect, named after Brian Josephson, who proposed it in the early 1960s while a graduate student at Cambridge University in the United Kingdom (he eventually received a Nobel Prize in Physics for it in 1973). To understand it, let us go through some basic facts about superconductors first.

Some materials (mainly metals), when cooled below a certain critical temperature (which varies from one material to another), loose all electrical resistivity (see Fig. 15.8), becoming perfect conductors ("superconductors").

This phenomenon is also accompanied by the "expulsion" of magnetic fields, namely the lack of penetration by external magnetic fields

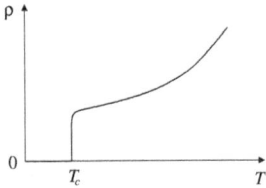

**Figure 15.8** Resistivity as a function of temperature for a superconducting material. $T_c$ is the critical temperature.

(the so-called Meissner effect). It turns out that conduction electrons in superconductors pair up, creating a coherent collective state that becomes immune to lattice defects and other mechanisms that scatter single electrons and cause finite resistivity. But this coherent superposition of paired electrons is relatively fragile; it needs low temperatures and can be destroyed if an applied magnetic field exceeds a certain magnitude. In standard metals such as aluminum, lead, niobium, and tin, the critical temperature $T_c$ varies between 1.2 and 9.3 kelvin; in some compounds containing rare-earth elements, $T_c$ can go as high as 125 kelvin at ambient pressure (these are the so-called high-temperature superconductors). Unfortunately, no one has been able to synthesize a compound that superconducts at room temperature *and* ambient pressure.[7]

When two superconductors are separated by a thin insulator, something very peculiar can happen: a dc (i.e., direct) current flows between them in the absence of applied bias voltage. The current comprises electron pairs (called Cooper pairs) that tunnel through the insulating barrier, as shown schematically in Fig. 15.9.

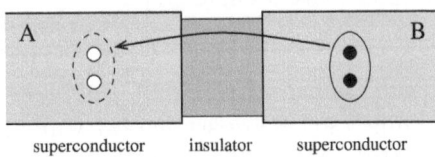

superconductor    insulator    superconductor

**Figure 15.9** Cooper pair tunneling through a Josephson junction barrier.

The electron pairs are the ones in the tunneling process; individual electrons do not participate.

---

[7]Superconductors with $T_c$ very close to room temperatures exist but only when submitted to impractical high pressures.

When a voltage bias is applied across the junction, an ac (i.e., alternating) current is generated.

It turns out that superconductors can be described by a single quantum probability amplitude,

$$\psi \approx \sqrt{n}\, e^{i\varphi},$$

where $n$ is the Cooper pair density and $\varphi$ is a phase. In a Josephson junction between superconductors $A$ and $B$ with individual phases $\varphi_A$ and $\varphi_B$, respectively, and a bias $V_{\text{bias}}$, the current is given by

$$I = I_0 \sin(\varphi)$$

where $\varphi = \varphi_A - \varphi_B$, with

$$\frac{d\varphi}{dt} = \frac{2e}{h} V_{\text{bias}}.$$

Here, $I_0$ is the so-called critical current and is a characteristic of the junction.

Since the Josephson junction is essentially a capacitor, if we denote the excess charge it stores as $Q$, we can write

$$Q = 2e\,(n_A - n_B) \times V,$$

with $V$ denoting the junction volume and $n_A$ and $n_B$ are the Cooper pair densities of each superconductor. Recalling that

$$I = -\frac{dQ}{dt} \quad \text{and} \quad Q = V_{\text{bias}}\, C_K,$$

where $C_J$ is the junction capacitance, we find

$$\frac{d\varphi}{dt} = \frac{2e}{\hbar}\frac{Q}{C_J} \quad \text{and} \quad \frac{dQ}{dt} = -I_o \sin(\varphi),$$

which describe the evolution of an oscillator. When $\varphi \ll 1$ (i.e., in the regime of small oscillations), we can write an approximate classical Hamiltonian function on the variables $Q$ and $\Phi = \varphi \times \Phi_0$, with $\Phi_0 = \hbar/2e$ (called flux quantum):

$$H_{JJ} = \frac{Q^2}{2C_J} + \frac{\Phi^2}{2L_J} + U_{\text{nl}}(\Phi), \tag{15.1}$$

where the constant $L_J = \Phi_0/I_0$ has the dimensions of an inductance.

The nonlinear potential $U_{nl}$ comprises $\Phi$-dependent terms beyond the quadratic form. This part is important and cannot be neglected (as far as qubit applications are concerned). When quantized, the quadratic, harmonic oscillator yields equally-spaced energy levels, making it impossible to distinguish transitions; the nonlinear potential breaks that and allows for each transition to be individually identified, as they correspond to emission or absorption of photons with different frequencies; see Fig. 15.10.

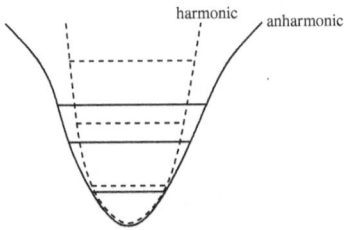

Figure 15.10 The effect of anharmonicity in a confining potential.

The two lowest-lying energy levels of a Josephson junction can be employed to define a qubit. However, there are various circuit implementations. Calling the Josephson junction a circuit element composed of a capacitor, an inductor, and a nonlinear barrier (see Fig. 15.11), we highlight a few fundamental implementations:

Figure 15.11 The Josephson junction circuit element.

- *Charge qubits*: see Fig. 15.12. In this setup one utilizes the two lowest states of the "Cooper pair box" as states $|0\rangle$ and $|1\rangle$. The anharmonicity created by the nonlinear potential allows these states to be accessed without populating other states. This kind of qubit was important in the early days of quantum computing but are too prone to decoherence and have been replaced by an alternative called *transmon* (see below).

- *Flux qubits (a.k.a. RF-SQUID qubit)*: see Fig. 15.13. In this setup, the variable $\Phi$ is modulated by a magnetic field flux $f$, such that $\phi \to \phi - (2\pi/\Phi_0) \times f$. Another term is added to the effective

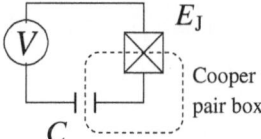

Figure 15.12 Circuit layout of a superconductor charge qubit.

Hamiltonian, resulting in a modulated "washboard" potential (see Fig. 15.14). There are multiple wells in the potential, each one corresponding to a different number of flux quanta. The two lowest-energy configurations are symmetric/antisymmetric pairs and are used as states $|0\rangle$ and $|1\rangle$.

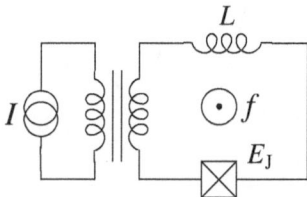

Figure 15.13 Circuit layout of a superconductor flux qubit.

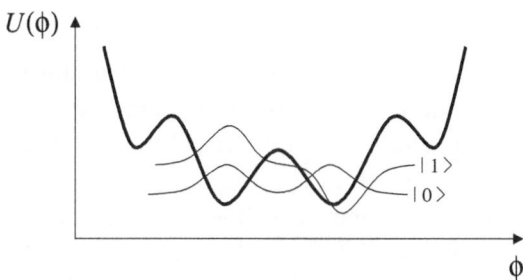

Figure 15.14 A modulated "washboard" potential (thick line) and its two lowest eigenstates (thin lines).

- *Phase qubits (a.k.a. current-biased qubits)*: see Fig. 15.15. In this setup, a current source is applied directly to the Josephson junction. The net result is to add a linear term to the "washboard" potential, see Fig. 15.16. The current can be adjusted to allow for only two states on a well (the other states leak out and are not stably populated); in fact, such a leakage can be exploited for measurement purposes.

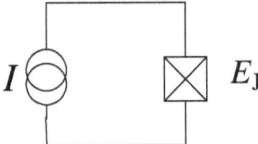

Figure 15.15  Circuit layout of a superconductor phase qubit.

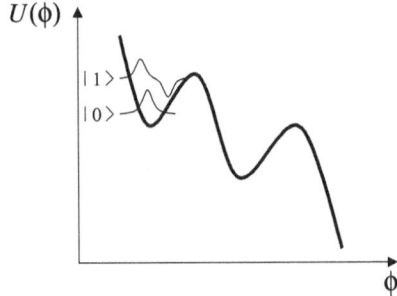

Figure 15.16  A tilted "washboard" potential (thick line) and the qubit states (thin lines).

There are many other implementations and architectures, such as transmons, Xmons, fluxoniums, quantoniums, etc. Most of them are hybrid and combine elements of those three fundamental setups. Transmons in particular are based on the Cooper pair box concept but have an added large shunting capacitance allowing $E_J$ to be much larger than the charging energy $E_C = e^2/2C_{tpt}$, where $C_{tot}$ is the total capacitance of the box. The biggest advantage is that the qubit energy spectrum becomes less dependent on the charge, reducing the qubit susceptibility to environmental electric field fluctuations (the so-called charge noise). This helps reduce decoherence. The drawback is that the states $|0\rangle$ and $|1\rangle$ become closer and the anharmonicity weaker, complicating the addressability of those states. Yet the treadoff is mostly positive and has lead to qubits of high quality. Google, IBM, and Rigetti have adopted the transmon architecture in their quantum processors.

Single-qubit gates are implemented with the help of transmission lines that resonate at frequencies $\omega$ corresponding to the energy separation between $|0\rangle$ and $|1\rangle$ states: $\omega = (E_1 - E_0)/\hbar$. Each individual qubit has its own attached transmission line and operate at a slightly different resonant frequency than the others. To perform two-qubit gates, qubits are physically coupled by tunable circuit elements such as capacitors and inductors; in this case, only nearest-neighbor (local) gates are possible. Alternatively, the qubits can be coupled to a cavity with quantized electromagnetic modes and these modes can be utilized to selectively couple pairs of qubits.

All operations in superconductor-based qubits use microwave pulses since $E_J$ and $E_C$ fall into that range of frequencies in the electromagnetic spectrum. As a result, these qubits can only operate at millikelvin temperatures (1 kelvin is approximately equivalent to 20 GHz).

Superconductor-based qubits benefited from decades of fundamental and applied research on Josephson junctions. They can also be readily fabricated with standard techniques (their sizes are in the micrometer range) and packed inside a chip. However, because of all the circuitry involved in their setups and the need to run microwave transmission lines and attached them to cavities, qubits are in close proximity and suffer from crosstalk. Moreover, the qubits are not true two-level systems and leakage can occur. It might be that they have already reached the lowest possible decoherence that nature affords them, in which case superconductor-based systems may never reach large-enough scales to run complex quantum algorithms. But they remain very popular and several companies are betting in this technology, not only for quantum computing but also other applications such as sensing and transduction.

## 15.4   DEFECT-BASED QUBITS

There is a variety of qubit realizations based on localized electronic states in solid-state systems, and they often originate from lattice defect. The defects can be naturally occurring or induced (e.g., via implantation of foreign atoms into a crystal, in which case the defective atom is named a dopant). They can involve a single atom in a crystal or multiple adjacent atoms. Some are addressable optically and/or via microwaves; others are fully electric or are a combination of electric and magnetic. Here we will only describe two of the most popular realizations: NV centers in diamond and charge donors (typically phosphorus) in silicon. They are paradigmatic in the sense that other defect-based qubits tend to follow the same physical principles of these two.

## 15.4.1 NV centers in diamond

Diamond is a particularly suitable substrate for qubits based on the electron's spin. It is an insulator with a very large band gap (5.47 eV), making it naturally immune to electronic (charge) excitations at room temperature. It is also transparent – the small amount of color one sees in some diamonds is actually due to lattice defects and dopants which are thus called color centers. Being transparent means that one can use light to access states located inside diamond. Finally, the predominant carbon isotope, $^{12}C$, has zero nuclear spin, leading to an absence of hyperfine coupling and suppression of this decoherence channel (at least to carbon nuclear spins).

How can one create localized spin states in diamond? It turns out that they occur naturally, but can also be induced by irradiation. In particular, the absence of a carbon atom (a lattice vacancy) next to a nitrogen dopant creates a color center known as NV. This defect induces electronic states suitable for a spin qubit.

The electrons bound to the NV center have total spin $S = 1$, inducing three sublevels with projection spin numbers $m = 0, +1, -1$; only $m = 0$ and $m = -1$ are used as qubit states. They can be accessed via microwave pulsing and show decoherence times ranging from microseconds to milliseconds. Most of the decoherence comes from coupling to the nuclear spins in the nitrogen dopant and from any $^{13}C$ isotopic contaminants. However, the nitrogen nuclear spins are actually a blessing, as they themselves can be considered super quiet qubits, with decoherence times close to 1 second! Transitions between the nuclear spin states, facilitated by the NV center electronic spin states, fall in the MHz range and can be easily accessed to create logic qubit systems.

Many techniques borrowed from quantum optics can be applied to NV centers to prepare, manipulate, and detect qubit states with very high fidelity.

The main challenge of NV center-based qubits is how to implement two-qubit gates. Two routes have been proposed: magnetic dipole-dipole interactions (qubit-qubit distances would need to be within the 10 nm range) and the use of optical waveguides to couple qubits coherently over larger distances. There have been some demonstrations of the latter where entanglement between two NV centers was achieved. Yet, it is unclear how this approach could be made to scale.

## 15.4.2 Phosphorus in silicon (P in Si)

Silicon is another material with a relatively large band gap and no nuclear spin in its predominant isotope ($^{28}$Si). The biggest advantage over diamond is the vast knowledge and experience in manipulating it at industrial scale, thanks to many decades of its use in electronics. For instance, one can selectively dope Si, including with $^{31}$P, an isotope of phosphorus that has a nuclear spin 1/2. The combination of the relatively inert silicon substrate with the $^{31}$P nuclear spin provides a localized qubit with very long decoherence times, from tens of milliseconds to seconds!

The original Si(P) qubit proposal called for P being a shallow dopant (i.e., near the Si surface) that could be controlled via electric gating, see Fig. 15.17. Recent advances in lithography and a more precise placement of P in Si allow deeper implantation.

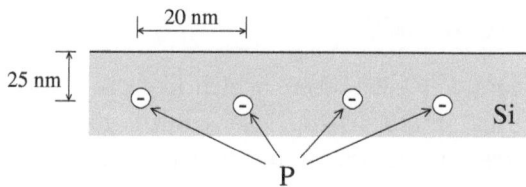

**Figure 15.17** Schematic of a Si(P) qubit system, with approximate average dimensions.

Unfortunately, it turns out that accessing directly the nuclear spin of P is very difficult. But one can readily access its excess valence electron; P effectively acts as a hydrogen-like atom embedded in a Si substrate. The valence electron can be removed from or added to P by electrostatic gating, and its spin used as a qubit. Moreover, it can be detected using a single-electron transistor (SET) placed nearby, see Fig. 15.18.

By monitoring the changes in the conductance of the SET, one can tell when the donor is charged or not with an extra valence electron. A combination of magnetic fields, microwave pulses, and current monitoring (in addition to gate pulsing) can be used to initialize, manipulate, and readout the electron's spin state.

Two-qubit operations can be performed by placing donors sufficiently close to each other, so that their electron spins can interact via exchange which can be controlled by electrostatic gates as well.

The main challenge in this approach is how to precisely control (within 1 nm) the location of implanted donors, or, alternatively, to

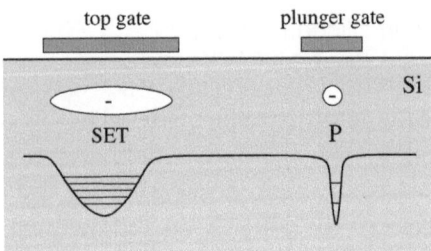

**Figure 15.18** Schematic of a Si(P) qubit setup with a SET. The potential energy profile is also shown.

couple donors that are far apart and when their exact locations are not known. Yet, much progress has been made in recent years, including the realization of small quantum processors.[8]

## 15.5  RYDBERG ATOMS

Finally, we discuss a qubit system which has progressed intensively in the last few years. It consists of laser-cooled, neutral atoms trapped by optical tweezers and magnetic means within a vacuum chamber, see Fig. 15.19. Arrays with hundreds of atoms are currently achievable. The main feature of these atoms (typically heavy alkali such as Rb or Cs), is that their valence electron is excited to a very large main quantum number $n$, putting them into states with very large radii, of the order of 1 $\mu$m.[9] A qubit can be encoded in the single-electron state via Zeeman splitting (i.e., using an external magnetic field), or through the hyperfine coupling to the nucleus spin. Both mechanisms produce doublets split by frequencies in the GHz range, which allows for their manipulation through microwave pulsing. It is also possible to employ all-electronic states for a qubit basis states, in which case manipulations can be performed with optical fields at the individual atom level. Similarly to ions in the ion-trapped-based qubits, qubit states in Rydberg neutral atoms also have nearby excited states which can be used for qubit initialization and readout.

---

[8]Thorvaldson, I., D. Poulos, C. M. Moehle, S. H. Misha, H. Edlbauer, J. Reiner, H. Geng, B. Voisin, M. T. Jones, M. B. Donnelly, L. F. Pena, C. D. Hill, C. R. Myers, J. G. Keizer, Y. Chung, S. K. Gorman, L. Kranz, and M. Y. Simmons. 2024. arXiv:2404.08741

[9]Recently, other alkali earth atoms such as Sr, which has two valence electrons, have also been employed due to some technical advantages.

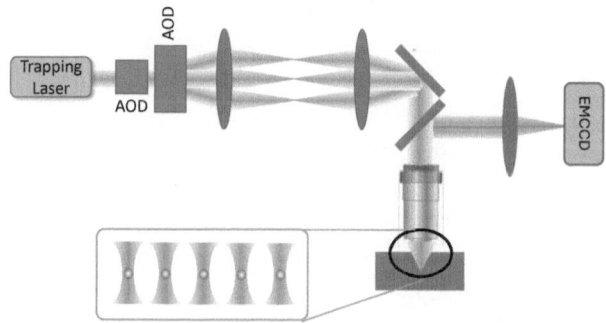

Figure 15.19 Schematic illustration of a setup for trapping Rydberg atoms in a regular array. AOD stands for acousto-optic deflector and EM-CCD for electron-multiplying charge-coupled device camera. Courtesy of Hebin Lin.

The intrinsic lifetime of large-$n$ Rydberg states due to spontaneous decay is very large due to the weak dipole coupling to low-$n$ states. Thus, in practice, the limiting factor in the lifetime of a Rydberg atom qubit is the coupling to external agents, such as radiation, confining fields, or non-trapped gaseous atoms. $T_2$ dephasing times of hundreds of ms can be achieved, but there are indications that they could be soon extended. Recent reports indicate a $T_2$ time of 20 s for single a Cs atom, and up to half a minute for Sr arrays. The physical trapping of the atom itself can last a very long time, reaching an hour.

The Rydberg atoms kept in a regular arrangement are typically a few $\mu$m apart. That allows for some small overlap of their orbital states and the appearance of van der Waals interactions. However, the prevalent form of interaction between Rydberg atoms in an array is dipole-dipole. This allows for a phenomenon called Rydberg blockade, where no two or more atoms within a certain radius cannot be simultaneously excited.

Single-qubit operations are performed via microwave pulsing but need to be combined with ways to spatially distinguish atoms for individual qubit addressability. This is usually done via magnetic-field gradients. An alternative is to use focused lasers, which can either induce energy shifts or populate auxiliary excited states that intermediate transitions between qubit states, as shown in Fig. 15.20.

Two-qubit operations explore the Rydberg blockade effect. For instance, a control-Z gate can be implemented by a sequence of pulses to the atoms involved when simultaneous occupancy of an intermediate

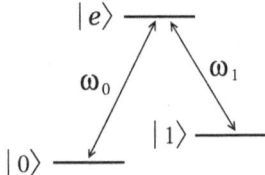

**Figure 15.20** A flip operation between $|0\rangle$ and $|1\rangle$ states in Rydberg atom qubit utilizing an intermediate excited atomic state $|e\rangle$ and two pulses of frequencies $\omega_0$ and $\omega_1$.

state in the atoms is forbidden. The scheme is shown in Fig. 15.21. An interesting characteristic of Rydberg atom-based systems is that they also allow for multi-qubit gates, which could be advantageous in terms of efficiency in implementing quantum algorithms.

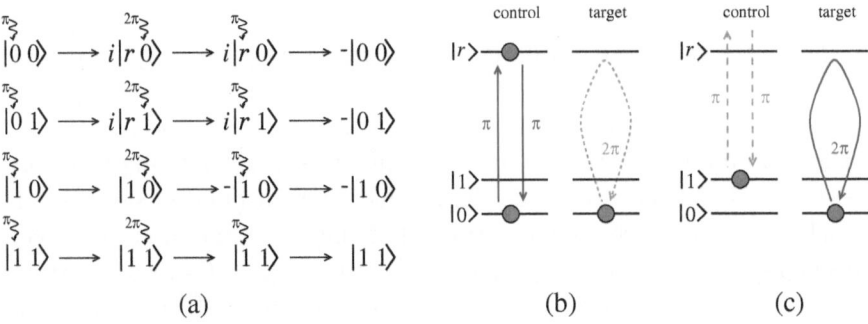

**Figure 15.21** (a) Pulse scheme for implementing a control-Z gate (up to an overall minus sign). The first qubit is the control and the second is the target. Three pulses are applied: a $\pi$ pulse to the control qubit, a $2\pi$ pulse on the target bit, and a second $\pi$ pulse on the control bit again. (b,c) Depending on the initial state, transitions involving the auxiliary state $|r\rangle$ are allowed or not due to detuning or the Rydberg blockade. Dashed lines indicate forbidden transitions.

In addition to implementing universal quantum computers, Rydberg atoms can be utilized for performing quantum simulations not only because they realize different array configurations but also for the tunability of inter-qubit interactions. In particular, they can be used to simulate spin-based magnetic systems of relevance to materials physics and whose properties are hard to compute with standard quantum computers.

There are still a few challenges to Rydberg atom quantum computing, but progress to overcome them has been steady. Loading the traps uniformly used to be very time consuming, with efficiencies as low as 50% per trap. But there are recent reports of loading efficiency reaching 90%. Another issue is scaling: arrays as large as $30 \times 30$ can be regularly obtained, but it is not obvious how to increase array sizes to many thousands of qubits, which will be likely necessary to implement quantum error correction at the level needed for running ambitious quantum algorithms such as Shor's and others. But progress on this front has been happening, with a recently reported 6,000-atom array being achieved, as well as a successful demonstration of moving the optical tweezers and shuttling atoms in the trapping region.

Rydberg quantum processors and quantum simulations are already commercially available.

## 15.6 PHOTONIC QUBITS

Photons are natural qubits since they intrinsically support two orthogonal polarization modes, see Fig. 15.22.[10] They can be easily and inexpensively produced and detected. They are also very resilient to decoherence and can travel very long distances. These qualities are balanced by an important caveat: photons interact very weakly, making it hard to implement multi-qubit operations. Their first use in quantum information processing was in the implementation of QKD protocols, where they serve as quantum channels (flying qubits). They are also important in quantum sensing. Their use for quantum computation is not as widespread as compared to other physical realizations but continues to evolve and improve.

While polarization is an obvious way to implement qubit states with photons, it is often not the most practical way.[11] Time bins, frequency, and spatial location can also be employed. Consider the example in Fig. 15.23, which shows time-bin qubit encoding: an incoming photon can take either of the two paths (short versus long one, typically with equal probability) after passing through a Mach-Zehnder interferometer (MZI).[12] Calling the short one "0" and the long one "1", the state of

---

[10]For instance, vertical $\times$ horizontal for linear polarization or right $\times$ left for circular polarization.

[11]Polarization can be unstable in optical fibers.

[12]A Mach-Zehnder interferometer consists of two beam splitters and two mirrors that create two optically identical paths for an incoming beam.

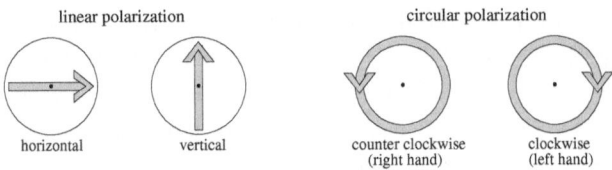

linear polarization       circular polarization

horizontal    vertical     counter clockwise    clockwise
(right hand)    (left hand)

Figure 15.22 Two types of light polarization modes: linear (static) and circular (rotating). Polarization is defined by the orientation of the electric field carried by the photon, which is always perpendicular to the direction of propagation. For each type of polarization, there are two orthogonal photon states.

the photon at the point where the two paths merge again is a superposition $|\psi\rangle = (|0\rangle + e^{i\phi}|1\rangle)/\sqrt{2}$. By changing the difference in length between the two paths, $\Delta l = l_1 - l_2$, one can vary the relative phase ($\phi = 2\pi\Delta l/\lambda \bmod 2\pi$, where $\lambda$ is the photon wavelength). One can measure the state of the qubit by recording the photon's time of arrival at a detector ("0" will arrive before "1"). The difference in arrival time between the two states is determined by the difference in path lengths: $\Delta t = \Delta l/c$.

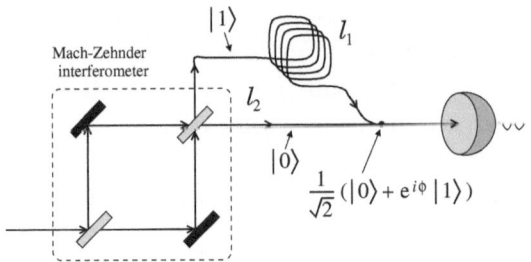

Figure 15.23 Scheme for creating a single-photon superposition state using time binning. A Mach-Zehnder interferometer is used to split the photon beam into two beams, with the beams taking path of different lengths before rejoining to create a superposition state.

Spatial encoding is another way to create a photonic qubit. In this case, it is the presence or absence of a photon that characterizes the basis states "0" and "1". The concept is illustrated in Fig. 15.24, where single- and dual-rail qubits are shown. In the single-rail encoding, one builds a superposition state through the presence or absence of a photon in a single mode. For dual-rail encoding, two spatially separate modes are employed and the superposition is between the two possible occupations for a single photon.

Dual rail can also be implemented more generally, with modes that are not necessarily spatially separated, e.g., two polarization modes colocated in the same optical fiber. The original proposal for LOQC mentioned in Ch. 12 employs dual-rail encoding. While simpler to implement, single-rail encoding does not allow for universal quantum computing when combined only with linear optical elements; dual rail does.

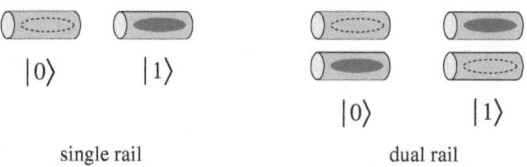

Figure 15.24 Spatial encoding of a single-photon qubit: single and dual rail using optical fibers. The empty empty and full ellipses represent unpopulated and populated photon modes, respectively.

Another way to encode quantum states with photons is through energy-time entanglement.[13] The concept originates from the Franson interferometer, which is illustrated in Fig. 15.25a: the monochromatic photons of an intense coherent light source pass through a nonlinear crystal and are "down converted" into a pair of photons.[14] Each one of these twin photons has an uncertain energy; however, the sum of their energy equals the energy of the incoming laser photon and is thus well defined.[15] Both photons are created at the same time (thus their ages are identical), but the exact moment of creation is uncertain. If we now split the photon path into two and direct each path toward a suitable interferometer with adjustable delays, we can create a Bell-state pair of the type $|00\rangle + |11\rangle$. Qubit states are encoded in the detection times.

We can simplify the energy-time entanglement setup to produce a single photon qubit suitable for implementation of the BB84 protocol, see Fig. 15.25b: replace the intense laser and the nonlinear crystal with a low-intensity light source, able to emit a single photon at a time, and

---

[13]Franson, J. D. 1989. *Bell inequality for position and time.* Phys. Rev. Lett. 62:2205-2208

[14]The yield of such a process is very low in practice, as low as 1 in $10^6$, and therefore most of the time a single photon emerges from the crystal for each incoming photon. Single-photon events are discarded.

[15]Recall that, for photons, energy and frequency are directly related: $E = \hbar f$.

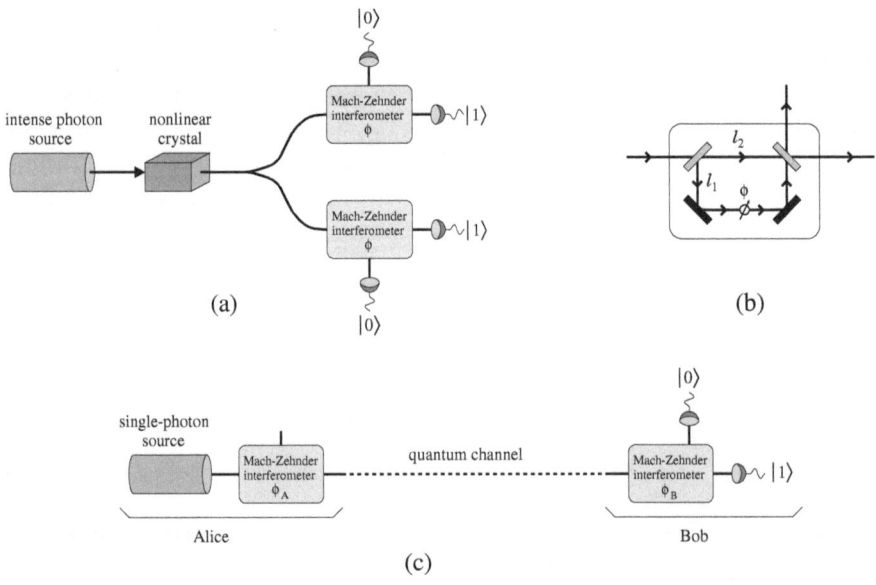

Figure 15.25 (a) A photonic Bell-state setup based on energy-time entanglement and the Franson interferometer. (b) The MZI used in (a) shown in more detail, with the long ($l_1$) and short ($l_2$) paths indicated, as well as the adjustable phase $\phi$. When both photons travel along the long arms or the short arms of their MZIs, they arrive simultaneously at the detectors; in this case, because each path is made equally probable, the Bell state $|00\rangle + |11\rangle$ is produced and detected. (c) A single-photon qubit implementation inspired by the setup in (a) and suitable for the BB84 protocol. The computational basis states correspond to photons arriving at one or the other detector.

two adjustable MZIs (one for Alice and another for Bob) to control the relative phase. In this case, the photon state will be a superposition of distinct arrival times. There are four possible scenarios, depending on the paths taken in the MZIs: short-short (SS), short-long (SL), long-short (LS), and long-long (LL). The SL and LS have the same time delay but are modulated by different phases ($\phi_B$ and $\phi_A$, respectively). This setup is much more stable than one relying on polarization encoding.

Finally, one can also employ frequency encoding using the dual-rail approach, with each mode corresponding to a distinct frequency.[16]

---

[16]Lukens, J. M. and P. Lougovski. 2017. *Frequency-encoded photonic qubits for scalable quantum information processing.* Optica 4:8-16

As mentioned earlier, regardless of the quantum state encoding adopted, for photonic qubits, multi-qubit gates are much harder to implement than single-qubit ones. In principle, a possible strategy is to employ a nonlinear medium: the passage of a photon alters some medium property, which in turn changes the state of another photon that is also passing through, creating a control-like two-qubit gate. Unfortunately, the magnitude of such an effect is exceedingly small, making it impractical. A better strategy is to employ ancillary photons, measurements, and protocols such as quantum teleportation. In this case, the "interaction" needed to realize a two-qubit gate results from the detection and measurement of ancillary photons.[17] Another strategy is to start with a multi-photon state which already carries some amount of entanglement, and perform only single-qubit operations and measurements. We have seen a few examples of these more efficient strategies in Ch. 12. They are all probabilistic, requiring multiple runs to reach the desired result within a preestablished accuracy. Recently, progress has been achieved with non-probabilistic two-qubit gates by employing hybrid systems that combine photonic qubits with atomic cavities (to enhance nonlinearity).[18]

## 15.7   REFERENCES AND FURTHER READING

1. Bruzewicz, C. D., J. Chiaverini, R. McConnell, and J. M. Sage. 2019. *Trapped-ion quantum computing: Progress and challenges.* Appl. Phys. 6:021314

2. Kjaergaard, M., M. E. Schwartz, J. Braumüller, P. Krantz, J. I.-J. Wang, S. Gustavsson, and W. D. Olivier. 2020. *Superconducting qubits: Current state of play.* Annu. Rev. Condens. Matter Phys. 11:369-395

3. Gisin, N. and R. Thew. 2007. *Quantum communication.* Nature Photon. 1:165-171

4. Weber, J. R., W. F. Koehl, J. B. Varley, A. Janotti, B. B. Buckley,

---

[17]An example is: Barz, S., I. Kassal, M. Ringbauer, Y. O. Lipp, B. Dakic, A. Aspuru-Guzik, and P. Walther. 2014. *A two-qubit photonic quantum processor and its application to solving systems of linear equations.* Sci. Rep. 4:6115

[18]Stolz, T., H. Hegels, M. Winter, B. Röher, Y.-F. Hsiao, L. Husel, G. Rempe, and S. Dürr. 2022. *Quantum-logic gate between two optical photons with an average efficiency above 40%.* Phys. Rev. X 12:021035

C. G. Van de Walle, and D. D. Awschalom. 2010. *Quantum computing with defects*. Proc. Natl. Acad. Sci. U.S.A. 107:8513-8518

5. McCallum, J. C., B. C. Johnson, T. Botzem. 2021. *Donor-based qubits for quantum computing in silicon*. Appl. Phys. Rev. 8:031314

6. Henriet, L., L. Beguin, A. Signoles, T. Lahaye, A. Browaeys, G.-O. Reymond, and C. Jurczak. 2020. *Quantum computing with neutral atoms*. Quantum 4: 327

7. Majidy, S., C. Wilson, and R. Laflamme. 2025. *Building Quantum Computers*. Cambridge Univ. Press. Chapters 4, 5, and 6.

# Practicalities

## A.1   MATRIX DIAGONALIZATION

Suppose we have an operator $\hat{A}$ with eigenvalues $\{a_n\}$ and corresponding eigenvectors $\{|a_n\rangle\}$, with $n = 1, \ldots, D$ and $D$ being the dimension of the space where $\hat{A}$ acts. We can then write

$$\hat{A}\,|a_n\rangle = a_n\,|a_n\rangle \tag{A.1}$$

for all $n$.

The process of finding the eigenvalues and eigenvectors of $\hat{A}$ is usually performed by first obtaining a matrix representation of $\hat{A}$. Let $\{|\phi_k\rangle\}_{k=1,\ldots,D}$ be a complete orthonormal basis in the space of dimension $D$ where $\hat{A}$ acts, namely,

$$\langle\phi_k|\phi_j\rangle = \delta_{k,j}. \tag{A.2}$$

The matrix elements of $\hat{A}$ are then given by

$$A_{kj} = \langle\phi_k|\hat{A}|\phi_j\rangle. \tag{A.3}$$

Since the basis is complete and orthonormal, we can decompose the eigenvectors as

$$
\begin{aligned}
|a_n\rangle &= \sum_{k=1}^{D}\langle\phi_k|a_n\rangle\,|\phi_k\rangle, \\
&= \sum_{k=1}^{D}\alpha_{kn}\,|\phi_k\rangle, 
\end{aligned}
\tag{A.4}
$$

where $\alpha_{kn} = \langle \phi_k | a_n \rangle$ is a scalar. Now, applying the operator $\hat{A}$ to both sides of Eq. (A.4), we obtain

$$\hat{A} |a_n\rangle = \sum_{k=1}^{D} \alpha_{kn} \hat{A} |\phi_k\rangle. \tag{A.5}$$

Combining Eqs. (A.1), (A.4), and (A.5), we obtain

$$\sum_{k=1}^{D} \alpha_{kn} \hat{A} |\phi_k\rangle = a_n \sum_{k=1}^{D} \alpha_{kn} |\phi_k\rangle. \tag{A.6}$$

Finally, contracting the ket-like Eq. (A.6) with a bra $\langle \phi_j |$ and employing Eq. (A.2), we obtain

$$\sum_{k=1}^{D} \alpha_{kn} A_{jk} = a_n \alpha_{jn}, \tag{A.7}$$

where $j$ can vary from 1 to $D$. The latter equation can be cast in a matrix form:

$$\begin{pmatrix} A_{11} & A_{12} & \cdots & A_{1D} \\ A_{21} & A_{22} & \cdots & A_{2D} \\ \vdots & \vdots & \ddots & \vdots \\ A_{D1} & A_{D2} & \cdots & A_{DD} \end{pmatrix} \begin{pmatrix} \alpha_{1n} \\ \alpha_{2n} \\ \vdots \\ \alpha_{Dn} \end{pmatrix} = a_n \begin{pmatrix} \alpha_{1n} \\ \alpha_{2n} \\ \vdots \\ \alpha_{Dn} \end{pmatrix}. \tag{A.8}$$

This form is likely familiar to those who have some experience with matrix algebra. We can take an extra step and incorporate the right-hand side term of Eq. (A.8) into its left-hand side after noticing that

$$a_n \begin{pmatrix} \alpha_{1n} \\ \alpha_{2n} \\ \vdots \\ \alpha_{Dn} \end{pmatrix} = \begin{pmatrix} a_n & 0 & \cdots & 0 \\ 0 & a_n & \cdots & 0 \\ \vdots & \vdots & \ddots & \vdots \\ 0 & 0 & \cdots & a_n \end{pmatrix} \begin{pmatrix} \alpha_{1n} \\ \alpha_{2n} \\ \vdots \\ \alpha_{Dn} \end{pmatrix}, \tag{A.9}$$

resulting in

$$\begin{pmatrix} A_{11} - a_n & A_{12} & \cdots & A_{1D} \\ A_{21} & A_{22} - a_n & \cdots & A_{2D} \\ \vdots & \vdots & \ddots & \vdots \\ A_{D1} & A_{D2} & \cdots & A_{DD} - a_n \end{pmatrix} \begin{pmatrix} \alpha_{1n} \\ \alpha_{2n} \\ \vdots \\ \alpha_{Dn} \end{pmatrix} = 0. \tag{A.10}$$

Equation (A.10) represents a homogenous system of $D$ coupled linear equations on the variables $\{\alpha_{jn}\}_{j=1,...,D}$. Therefore, to have a nontrivial solution of the system equation, the determinant of the matrix part must be zero, namely,

$$\det \begin{pmatrix} A_{11} - a_n & A_{12} & \cdots & A_{1D} \\ A_{21} & A_{22} - a_n & \cdots & A_{2D} \\ \vdots & \vdots & \ddots & \vdots \\ A_{D1} & A_{D2} & \cdots & A_{DD} - a_n \end{pmatrix} = 0. \qquad (A.11)$$

The determinant of the matrix part is equal to a polynomial of order $D$ on the unknown variable $a_n$. Equation (A.11) says that the roots of this polynomial are the sought eigenvalues.

Consider an operator $\hat{A}$ acting on $D = 3$ space with the following matrix form:

$$\hat{A} = \begin{pmatrix} 2 & \sqrt{2} & 0 \\ \sqrt{2} & 2 & \sqrt{2} \\ 0 & \sqrt{2} & 2 \end{pmatrix}$$

(notice that the matrix is Hermitian and therefore we expect all eigenvalues to be real and the eigenvectors to form a complete orthonormal set). Let us call $a$ the eigenvalue variable, in which case we can write

$$\det \begin{pmatrix} 2 - a & \sqrt{2} & 0 \\ \sqrt{2} & 2 - a & \sqrt{2} \\ 0 & \sqrt{2} & 2 - a \end{pmatrix} = 0.$$

Applying standard rules for computing determinants, we obtain

$$(2 - a)^3 - 4(2 - a) = 0. \qquad (A.12)$$

Solving for $a$, we obtain the three eigenvalues $a_1 = 0$, $a_2 = 2$, and $a_3 = 4$.

What about the eigenvectors corresponding to these eigenvalues? A simple way to get them is to substitute the eigenvalue back into the matrix in Eq. (A.10) and use all but one of the coupled equations to find the amplitudes $\alpha_{jn}$.

For instance, let us determine the eigenvector corresponding to $a_1 = 0$. The following set of equations need to be satified:

$$2\alpha_{11} + \sqrt{2}\alpha_{21} = 0$$
$$\sqrt{2}\alpha_{11} + 2\alpha_{21} + \sqrt{2}\alpha_{31} = 0$$

resulting in

$$\alpha_{21} = -\sqrt{2}\,\alpha_{11}$$
$$\alpha_{31} = \alpha_{11}.$$

To find a value for $\alpha_{11}$, we impose that the eigenvector is normalized, which you can easily show to correspond to the condition

$$|\alpha_{11}|^2 + |\alpha_{21}|^2 + |\alpha_{31}|^2 = 1,$$

which in this case results in $|\alpha_{11}| = 1/2$. Since this is the first eigenvector we computed, we are free to set it real and positive and can choose $\alpha_{11} = 1/2$.

What about the other eigenvectors? We follow the same procedure. Consider the second eigenvalue, $a_2 = 2$. Substituting it in Eq. (A.10) and this time dropping the first equation, we get the set of coupled equations

$$\sqrt{2}\,\alpha_{12} + \sqrt{2}\,\alpha_{32} = 0$$
$$\sqrt{2}\,\alpha_{22} = 0$$

resulting in $\alpha_{22} = 0$ and $\alpha_{32} = -\alpha_{12}$. Imposing normalization, we find $|\alpha_{12}| = 1/\sqrt{2}$ and pick $\alpha_{12} = 1/\sqrt{2}$. Notice that this eigenvector the second is orthogonal the first one we found, as expected. Finally, to obtain the third and last eigenvector, we can either substitute $a_3$ in Eq. (A.10) and follow the same procedure used for the two other eigenvectors, or notice the following: since the three eigenvectors form a complete orthonormal set, the third one must be orthogonal to the other two and have norm 1. Except for an overall phase factor, these conditions complete restrict which vector can be the third eigenvector. It is easy to verify that the amplitudes $\alpha_{13} = \alpha_{33} = 1/2$ and $\alpha_{23} = 1/\sqrt{2}$ satisfy these conditions and correspond to the eigenvalue $a_3$.

In practice, one hardly ever tries to find by hand the eigenvalues and eigenvectors of matrices larger than $2 \times 2$. Unless the matrices have symmetries, the polynomial root equations can be tough to solve even for $3 \times 3$ matrices and possibly impossible for matrices larger than $4 \times 4$. There are other methods based on similarity transformations and reduction to a tridiagonal form which are more suitable for large matrices and can be easily coded if numerical solutions are required. Packages and libraries such as NumPy (for Python) and LAPACK (for fortran) offer various methods for computing eigenvalues and eigenvectors of a matrix.

## A.2 PROJECTORS

A projector is an operator that extracts from a state vevtor its projection onto a particular subspace. Let us illustrate this concept through examples. Suppose you have a two-qubit state

$$|\psi\rangle = \alpha |01\rangle + \beta |11\rangle$$

with $|\alpha|^2 + |\beta|^2 = 1$. If you want to determine the probability of measuring the second qubit and obtaining 1, you use the following projector:

$$\hat{P} = \hat{I}_1 \otimes (|1\rangle\langle 1|)_2 \,,$$

where the subscript 1 signifies that the operator acts on the subspace of the first qubit while the subscript 2 does similarly for the second qubit. Then,

$$p(\text{2nd qubit } = 1) = \langle \psi | \hat{P} | \psi \rangle.$$

In words: the probability is equal to the expectation value of the appropriate projector. In this case,

$$
\begin{aligned}
p(\text{2nd qubit} = 1) &= \left( \alpha^* \langle 01| + \beta^* \right) \left( \hat{I}_1 \otimes (|1\rangle\langle 1|)_2 \right) \left( \alpha|01\rangle + \beta|11\rangle \right) \\
&= \left( \alpha^* \langle 01| + \beta^* \langle 11| \right) \left( \alpha|01\rangle + \beta|11\rangle \right) \\
&= |\alpha|^2 + |\beta|^2 \\
&= 1.
\end{aligned}
$$

Let us now ask: what is the probability of measuring the first qubit and obtaining 0? To answer this question, we use the projector

$$\hat{P} = (|0\rangle\langle 0|)_1 \otimes \hat{I}_2$$

and compute

$$
\begin{aligned}
p(\text{1st qubit} = 1) &= \langle \psi | \hat{P} | \psi \rangle \\
&= \left( \alpha^* \langle 01| + \beta^* \right) \left( (|0\rangle\langle 0|)_2 \otimes \hat{I}_2 \right) \left( \alpha|01\rangle + \beta|11\rangle \right) \\
&= \left( \alpha^* \langle 01| + \beta^* \langle 11| \right) \left( \alpha|01\rangle \right) \\
&= |\alpha|^2.
\end{aligned}
$$

## A.3   TENSOR PRODUCTS OF OPERATORS

We often encounter situations where we need to multiply two or more operators acting on different qubits or subspaces:

$$\hat{O} = \hat{A} \otimes \hat{B} \otimes \hat{C} \otimes \dots .$$

We call such an operation a tensor product. Let us use a tensor product of two single-qubit operators to illustrate how to implement this operation. We assume that the single-qubit operators are represented in their computational bases, namely,

$$\hat{A} = \begin{pmatrix} a_{00} & a_{01} \\ a_{10} & a_{11} \end{pmatrix} \quad \text{and} \quad \hat{B} = \begin{pmatrix} b_{00} & b_{01} \\ b_{10} & b_{11} \end{pmatrix}.$$

The tensor product is then given by

$$\hat{A} \otimes \hat{B} = \begin{pmatrix} a_{00} \begin{pmatrix} b_{00} & b_{01} \\ b_{10} & b_{11} \end{pmatrix} & a_{01} \begin{pmatrix} b_{00} & b_{01} \\ b_{10} & b_{11} \end{pmatrix} \\ a_{10} \begin{pmatrix} b_{00} & b_{01} \\ b_{10} & b_{11} \end{pmatrix} & a_{11} \begin{pmatrix} b_{00} & b_{01} \\ b_{01} & b_{11} \end{pmatrix} \end{pmatrix}$$

$$= \begin{pmatrix} a_{00} b_{00} & a_{00} b_{01} & a_{01} b_{00} & a_{01} b_{01} \\ a_{00} b_{10} & a_{00} b_{11} & a_{01} b_{10} & a_{01} b_{11} \\ a_{10} b_{00} & a_{10} b_{01} & a_{11} b_{00} & a_{11} b_{01} \\ a_{10} b_{10} & a_{10} b_{11} & a_{11} b_{10} & a_{11} b_{11} \end{pmatrix}.$$

It is straightforward to generalize this implementation to operators involving subspaces with dimensions larger than 2.

Notice that the dimensionality of the operator resulting from a tensor product is always equal to the product of the dimensions of the operators involved. In this case, because $\dim(\hat{A}) = 2$ and $\dim(\hat{B}) = 2$, $\dim(\hat{A} \otimes \hat{B}) = \dim(\hat{A}) \dim(\hat{B}) = 4$.

In the case of a tensor product involving more than two operators, we implement the product by parts. For instance,

$$\hat{O} = \hat{A} \otimes \left( \hat{B} \otimes \hat{C} \right).$$

Here, we first carry out the tensor product of $\hat{B}$ and $\hat{C}$ (call it $\hat{O}_{BC}$), and then the tensor product between $\hat{A}$ and $\hat{O}_{BC}$.

It is important to notice that while operators in a tensor product commute (since they act on different subspaces), the implementation of a tensor product in a matrix representation requires a fixed order, and the same order needs to be employed when taking the tensor product of corresponding state vectors. To understand this point, consider again the example of the two single-qubit operators $\hat{A}$ and $\hat{B}$ presented above. Let $|\psi_A\rangle$ and $|\psi_B\rangle$ and be single-qubit state vectors in the subspaces of $\hat{A}$ and $\hat{B}$, respectively. In their computational basis, we can write

$$|\psi_A\rangle = \alpha_0|0\rangle_A + \alpha_1|1\rangle_A = \begin{pmatrix} \alpha_0 \\ \alpha_1 \end{pmatrix}$$

and

$$|\psi_B\rangle = \beta_0|0\rangle_B + \beta_1|1\rangle_B = \begin{pmatrix} \beta_0 \\ \beta_1 \end{pmatrix}.$$

The tensor product of these two single-qubit state vectors can be straightforwardly computed with basis kets:

$$\begin{aligned} |\psi_A\rangle \otimes |\psi_B\rangle &= (\alpha_1|0\rangle_A + \alpha_2|1\rangle_A) \otimes (\beta_1|0\rangle_B + \beta_2|1\rangle_B) \\ &= \alpha_0\beta_0|0\rangle_A|0\rangle_B + \alpha_0\beta_1|0\rangle_A|1\rangle_B + \alpha_1\beta_0|1\rangle_A|0\rangle_B \\ &\quad + \alpha_1\beta_1|1\rangle_A|1\rangle_B. \end{aligned}$$

The proper tensor product in a matrix representation, compatible with that used for $\hat{A} \otimes \hat{B}$, is then equal to

$$|\psi_A\rangle \otimes |\psi_B\rangle = \begin{pmatrix} \alpha_0 \begin{pmatrix} \beta_0 \\ \beta_1 \end{pmatrix} \\ \alpha_1 \begin{pmatrix} \beta_0 \\ \beta_1 \end{pmatrix} \end{pmatrix} = \begin{pmatrix} \alpha_0\beta_1 \\ \alpha_0\beta_1 \\ \alpha_1\beta_0 \\ \alpha_1\beta_1 \end{pmatrix}.$$

## A.4 SOLVING A LINEAR MATRIX DIFFERENTIAL EQUATION

Let $M(t)$ represent a time-dependent, $N \times N$ square matrix satisfying the first-order differential equation

$$\frac{d}{dt}M(t) = \hat{\Lambda} M(t), \tag{A.13}$$

where $\hat{\Lambda}$ is a linear superoperator, namely, a linear operator that acts on matrices. When $\hat{\Lambda}$ is time independent and given some initial matrix $M(0)$, this equation can be formally solved:

$$M(t) = e^{\hat{\Lambda}t} M(0). \tag{A.14}$$

In practice, without exploring any symmetries or simplifications that may exist in $\hat{\Lambda}$, we deal with the exponentiation of a superoperator by employing diagonalization. Let us start by introducing indices in Eq. (A.13):

$$\frac{dM_{i,j}}{dt} = \sum_{k,l=1}^{N} \Lambda_{i,j;k,l} \, M_{k,l}. \tag{A.15}$$

Next, let us introduce single indices $I$ and $K$ that run over all values of the indices $(i,j)$ and $(k,l)$, respectively. One way to do that is to write

$$\begin{aligned} I &= (i-1)N + j \\ K &= (k-1)N + l, \end{aligned}$$

for $i,j,k,l = 1, \ldots, N$. Notice that both $I$ and $J$ run over $N^2$ values. We can then recast Eq. (A.15) as

$$\frac{dM_I}{dt} = \sum_{K=1}^{N^2} \Lambda_{I,K} \, M_K. \tag{A.16}$$

The matrix $M$ is now a vector of length $N^2$ and the superoperator $\Lambda$ is a $N^2 \times N^2$ matrix. When this matrix is diagonalizable, we can decompose it as

$$\hat{\Lambda} = \hat{V} \hat{\lambda} \hat{V}^{-1},$$

where $\hat{\lambda}$ is a diagonal matrix and $\hat{V}$ is some $N^2 \times N^2$ invertible square matrix. Then, we can rewrite Eq. (A.16) as

$$\frac{d\tilde{M}_I}{dt} = \lambda_I \, \tilde{M}_I, \tag{A.17}$$

where $\tilde{M} = \hat{V}^{-1} M$. The differential equation (A.17) is diagonal and can be readily solved element-by-element to yield

$$\tilde{M}_I(t) = e^{\lambda_I t} \, \tilde{M}_I(0), \tag{A.18}$$

which can be rewritten in terms of the vector components of the original matrix $M$ as

$$M_I(t) = \sum_{K=1}^{N^2} \left[ \hat{V} \, e^{\hat{\lambda} t} \, \hat{V}^{-1} \right]_{I,K} M_K(0).$$

Comparing this equation to Eq. (A.14), we identify

$$e^{\hat{\Lambda} t} = \hat{V} \, e^{\hat{\lambda} t} \, \hat{V}^{-1}. \tag{A.19}$$

# Qiskit

## B.1  INSTALLATION

Qiskit is constantly evolving and therefore we recommend that the reader checks the installation instructions provided in the package distribution website, which is currently found at

https://docs.quantum.ibm.com/guides/install-qiskit

Here, we list the main steps found on that website at the time when the book was going to press (Qiskit version 1.0).

1. If you do not already have it, install the latest version of Python (the programming language) on your computer:

    https://www.python.org/

2. If pip is not included with your Python installation, install it:

    https://pip.pypa.io/en/stable/installation/

3. Create a directory in your computer to run your qiskit codes; we suggest naming it "qip-qiskit". Open a shell command window and go to that directory.

4. Within that directory, create a Python virtual environment through the commands:

    – for Mac or Linux:

    ```
    python3 -m venv /path-to-qip-qiskit
    source /path-to-qip-qiskit/bin/activate
    ```

    – for Windows:

```
python3 -m venv c:\path-to-qip-qiskit
c:\path-to-qip-qiskit\Scrips\Activate.ps1
```

where "path-to-qip-qiskit" is the path in your system to the directory qip-qiskit.

5. Install Qiskit and a couple more useful packages through the commands:

```
pip install qiskit[visualization]
pip install qiskit-aer
pip install qiskit-ibm-runtime
```

6. Use the command line to install JupyterLab for composing and executing Qiskit codes:

```
pip install jupyter
```

7. Now you are ready to start a jupyter session and create/open notebooks. Enter the command

```
jupyter notebook
```

## B.2 KEY COMMANDS

- Loading common qiskit modules:

```
from qiskit import QuantumCircuit
from qiskit.quantum_info import Statevector
from qiskit.quantum_info import Operator
from qiskit.primitives import StatevectorSampler
from qiskit.visualization import plot_histogram
from qiskit.visualization import plot_state_city
from qiskit.visualization import plot_bloch_vector
from qiskit.visualization import \
plot_bloch_multivector
```

- Create a $n$-qubit circuit object (qubits set to $|0\rangle$ by default):

```
circuit = QuantumCircuit(n)
```

- Apply a Hadamard gate to qubit 0:

```
circuit.h(0)
```

- Apply an $R_x(\theta)$ gate to qubit 0, where $\theta$ is the angle:

```
circuit.rx(theta,0)
```

- Apply an $R_y(\theta)$ gate to qubit 0, where $\theta$ is the angle:

```
circuit.ry(theta,0)
```

- Apply a phase gate to qubit 9 with a phase $\phi$:

```
circuit.p(phi)
```

- Apply a CNOT gate to qubits 0 (control) and 1 (target):

```
circuit.cx(0,1)
```

- Apply a control phase gate on qubits 0 and 1 with a phase $\phi$:

```
circuit.cp(phi, 0, 1)
```

- Obtain and print the state vector at the current output of a circuit:

```
state = Statevector(circuit)
print(state)
```

- Draw a circuit using MatPlotlib:

```
circuit.draw("mpl")
```

- Plot a state vector on the Bloch sphere given Cartesian coordinates $(x, y, z)$:

```
plot_bloch_vector([x,y,z])
```

- Plot a state vector on the Bloch sphere given spherical coordinates $(r, \theta, \phi)$:

```
plot_bloch_vector([r,theta,phi]),
coord_type = 'spherical')
```

- Plot a state vector on the Bloch sphere

```
plot_bloch_multivector(state)
```

- Apply measurements to all qubits in the circuit:

```
circuit.measure_all()
```

- Run a sampling evaluator $n$ times and retrieve results:

```
sampler = StatevectorSampler(default_shots=n)
job = sample.run([circuit])
result = job.results()
```

- Extract counts from measurements and print them:

```
data = result[0].data
counts = data.meas.get_count()
print(f"Counts: counts")
```

- Plot a histogram of counts:

```
plot_histogram(counts)
```

- Plot the elements of the density matrix given the state vector:

```
plot_state_city(state)
```

- Retrieve the unitary operator represented by a circuit and print it:

```
op = Operator.from_circuit(circuit)
op.draw("latex")
```

# Complexity Classes

Computational problems are classified according to their type and complexity. Some problems are of the decision type, when one is asked whether a solution exists or not. Others are of the determination type, where one is asked to find at least one solution if it exists. There are also problems of the counting type, where one is asked how many solutions exist.

As far as complexity, classical decision problems have four main classes. The time scaling is relative to the size of the problem (i.e., number of variables involved).

- **P** (polynomial time) – decision problems solvable by a standard, deterministic processor in polynomial time.

- **NP** (nondeterministic polynomial time) – decision problems for which there is a known deterministic polynomial-time algorithm to verify a "yes" answer.

- **NP-hard** – problems to which every NP problem can be reduced to in polynomial time.

- **NP-complete** – problems that are in the NP and in the NP-hard classes.

According to these definitions, if one could solve any problem in the NP-hard class in polynomial time, then all problems in NP would also be solvable in polynomial time. In this case, P = NP. This is widely believed to be impossible, implying P≠NP, although there is no known rigorous proof. An example of an NP problem is factoring: it takes no more than a polynomial-size computation to verify if a given set of numbers $\{X, Y, \ldots\}$ factorizes another one, $Z$ (all you need to perform multiplications and check if $X \cdot Y \cdots = Z$). But we do not know any classical algorithm that finds the factors in polynomial time.

Examples of NP-complete problems are the traveling salesman, 3-SAT, sudoku, battleship, graph coloring, integer programming, and knapsack. All NP-complete problems are also NP-hard, but the reverse is not true. An example of the latter is the famous halting problem in computer science ("given a program and its input, will it run forever?").

There are also complexity classes that invoke quantum computations. A very important one is **BQP** (bounded-error quantum polynomial time) – decision problems that can be solved in polynomial time by a quantum computer. Since quantum computers can simulate classical ones efficiently (i.e., with a polynomial overhead), it is known that P⊆BQP. However, the reverse is not known. For instance, factoring is in BQP (because of Shor's algorithm), but it is not known to be in P.

Currently, it is not known if NP⊆BQP. We do not know if all problems in NP can be solved in polynomial time by a quantum computer. In fact, this is a big open question in quantum complexity theory. We do not even known the answer to the reverse question: is BQP⊆NP? It is plausible that there problems solvable by a quantum computer whose answer cannot be easily verified classically, but a rigorous proof is still lacking.

In addition to time, one can also consider other resources needed to solve a computational problem. An obvious one is space, which we can roughly translate as the amount of "memory" needed. In this case, an important complexity class emerges: **PSPACE** (polynomial space) – problems solvable in a digital computer using a polynomial amount of memory but possibly an exponential amount of time. P is included in PSPACE, and so is NP and BQP (the latter has been proved by Bernstein and Vazirani).[1] Thus, our best bounds for BQP at the moment is P⊆BQP⊆PSPACE. As a result, we do not know, for instance, if factoring is not in P. Any proof of P≠BQP requires P≠PSPACE, which is an unsolved problem on its own. Our most current understanding of these complexity classes is illustrated in Fig. C.1.

Finally, there is another complexity class known as **QMA** (quantum Merlin Arthur) that is analogous to NP but uses a quantum computer to

---

[1]Bernstein, E. and U. Vazirani. 1997. *Quantum complexity theory*. SIAM J. Comput. 26:141101473

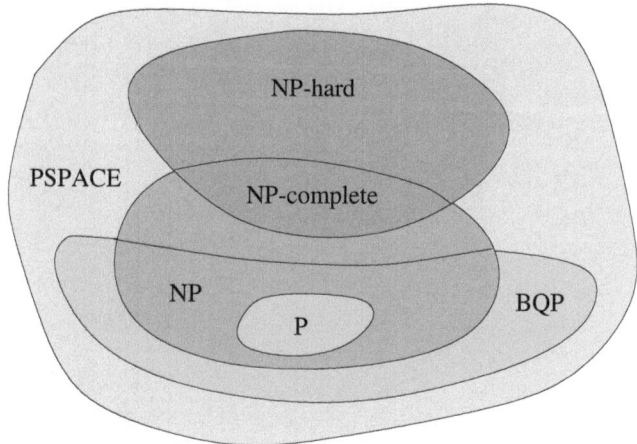

Figure C.1  Relations among various complexity classes as currently known or suspected.

verify the validity of the solution.[2] QMA contains BQP, as well as P and NP. But it is larger than NP and BQP. An example of a problem in QMA that is outside both NP and BQP is whether a $n$-qubit Hamiltonian with local interactions has an eigenvalue smaller than $a$ or if all eigenvalues are larger than $b$, where $b - a \geq O(1/n)$. QMA is contained in PSPACE.

---

[2]Since the verification uses a quantum computer, it is always probabilistic.

# Index